ON COLD IRON

A STORY OF HUBRIS
AND THE 1907 QUEBEC BRIDGE COLLAPSE

DAN LEVERT

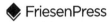

Suite 300 - 990 Fort St
Victoria, BC, V8V 3K2
Canada

www.friesenpress.com

Copyright © 2020 by Dan Levert
First Edition — 2020

Photos courtesy of Library and Archives Canada & the Kanien'kehaka Raotitiohkwa Cultural Center

Ritual of the Calling of an Engineer, University of Toronto Archives, Professor H.E.T. Haultain- fonds

Charles Carrington, Rudyard Kipling, His Life and Work, (London: Macmillan & Co. Ltd., 1955)

William D. Middleton, The Bridge at Quebec (Indiana University Press, 2001)

Royal Commission Report, Quebec Bridge Inquiry, Sessional Papers, No. 154 (1908)

Transcript of the Inquest, Bibliothèque et Archives Nationales du Québec (BANQ)

Prof. Eda Kranakis (U. of Ott.), Fixing the Blame, Organizational Culture and the Quebec Bridge Collapse, The International Quarterly of the Society for the History of Technology

All rights reserved.
No part of this publication may be reproduced in any form, or by any means, electronic or mechanical, including photocopying, recording, or any information browsing, storage, or retrieval system, without permission in writing from FriesenPress.

ISBN
978-1-5255-6220-4 (Hardcover)
978-1-5255-6221-1 (Paperback)
978-1-5255-6222-8 (eBook)

1. HISTORY / CANADA / POST-CONFEDERATION
2. TECHNOLOGY & ENGINEERING / HISTORY
3. TECHNOLOGY & ENGINEERING / CIVIL / BRIDGES

Distributed to the trade by The Ingram Book Company

TABLE OF CONTENTS

1	**PROLOGUE**		109	**CHAPTER 14**
				Thursday, August 29th
3	**INTRODUCTION**			
			118	**CHAPTER 15**
5	**CHAPTER 1**			*The Collapse*
	Professor Haultain's Request			
			145	**CHAPTER 16**
11	**CHAPTER 2**			*The Devastation*
	Rudyard Kipling's Response			
			154	**CHAPTER 17**
21	**CHAPTER 3**			*The Coroner's Inquest*
	A Bridge at Quebec			
			168	**CHAPTER 18**
29	**CHAPTER 4**			*The Royal Commission of Inquiry*
	Chief Engineer			
			179	**CHAPTER 19**
39	**CHAPTER 5**			*The Inquest Ends*
	Specifications and Financing			
			184	**CHAPTER 20**
49	**CHAPTER 6**			*The Inquiry Continues*
	Design			
			199	**CHAPTER 21**
60	**CHAPTER 7**			*The Inquiry Ends*
	Inspection Regime			
			214	**CHAPTER 22**
68	**CHAPTER 8**			*The Findings of the Commission*
	The Builders			
			235	**CHAPTER 23**
75	**CHAPTER 9**			*The Aftermath*
	The Dropped Chord A9-L			
			249	**EPILOGUE**
81	**CHAPTER 10**			
	The 1905 & 1906 Seasons		251	**GLOSSARY**
86	**CHAPTER 11**		258	**NOTES**
	1907 Early Warning Signs			
96	**CHAPTER 12**			
	Tuesday, August 27th			
102	**CHAPTER 13**			
	Wednesday, August 28th			

To my wife Sue, to Mom, my sons Brett and Phil, and in memory of Kate.

PROLOGUE

IN THE EARLY EVENING OF December 4, 1978, I sat in my apartment, drinking a glass of twelve-year-old scotch. I was celebrating. Tonight, I would receive my Iron Ring. I had completed the requirements for a Bachelor of Science Degree in Civil Engineering from the University of New Brunswick in Fredericton, and I had a job lined up with a major construction company to build a dam and hydroelectric facility in James Bay, Quebec. Life was grand.

Later that evening, a small group of students gathered with senior professional engineers in a modest, dimly lit room in downtown Fredericton, away from the university campus. My structures professor, a professional engineer and consultant to private industry, was present along with several other prominent engineers, one of whom was to lead the ceremony titled "The Ritual of the Calling of an Engineer." I was a bit surprised when the senior supervising engineer got up and locked the door; I think that's when I started to take this seriously. Two hours later the Obligation Ceremony and dinner were over; I had taken my Obligation and the Iron Ring circled the small finger of my working hand where it had been placed by Professor Francis. When the senior engineer unlocked the door, I remember feeling as though I'd been under a spell. The Ceremony had been solemn. I had taken part in something memorable. I felt as though I belonged here with these people and I didn't want to leave. I felt nurtured. The Ceremony had changed my view of engineering, given me a new and different perspective on my chosen profession. That being said, my classmates and I retired to a local pub to celebrate. I think we all felt the same though; the celebration was subdued.

It would be many years before I would come to fully appreciate the meaning and significance of the Obligation and its symbol. Although my Obligation hangs on my office wall, I have to keep reminding myself that my ring is there as a symbol of my Obligation, that it is not a sign to the world that I am a Canadian engineer. This book is my attempt to remind myself of the Obligation I took that night more than forty years ago, and its true meaning. I hope it will also serve as a reminder to my fellow Obligated engineers.

INTRODUCTION

THIS BOOK TELLS TWO STORIES. The first one is about the Ritual of the Calling of an Engineer and the events that led to the first Obligation Ceremonies in 1925. Near the time when a Canadian engineering student completes his or her degree requirements, they are invited to take part in the Ritual. During the Ceremony they take an Obligation that binds them to certain principles and ideals to guide them throughout their careers. You will be introduced to the two people who conceived the idea of the Ritual: Professor Herbert E.T. Haultain, of the University of Toronto, and Rudyard Kipling, the well-known British author and Nobel Laureate. Events in 1923 conspired to bring these two together at a critical moment in the history of the engineering profession in Canada. In 2025 the Ritual will be celebrating its 100th anniversary. The Iron Ring continues as a symbol for Canadian engineers, but more importantly, as a reminder to the wearer of his or her Obligation. Regarding the Ritual itself, it is "neither for the public nor the press." Obligated engineers are encouraged, however, to frame and mount their Obligation on their office wall, as I have.

The second story is about the collapse of the Quebec Bridge on August 29, 1907. There is a myth that prevails that the early Iron Rings were made from material salvaged from the steel wreckage of that bridge. There is no evidence of such a physical link. There is, however, an ethical link. The Obligation taken by the engineering graduate during the Ceremony acknowledges one's own assured failures and derelictions; it recognizes human frailty and teaches humility. There was no trace of humility in any of the senior engineers involved with the design of the Quebec Bridge. Their arrogance and absolute confidence in their work prevented them from realizing what the workers building the bridge already knew: that the bridge was failing under its own weight. Although the bridge could not have been saved, had there been even a trace of humility, the seventy-six men who lost their lives might have been spared. The lesson learned from the collapse of the Quebec Bridge is what we as engineers are expected to learn from Rudyard Kipling's Ritual: the lesson of humility.

Following the collapse of the Quebec Bridge, the coroner for the province of Quebec held an inquest into the death of the workers. The coroner's jury failed to discover the cause of the collapse. The day after the collapse, the federal government appointed a Royal Commission of Inquiry comprising three prominent Canadian engineers to inquire into the cause of the collapse. The commissioners spent months examining witnesses in Canada and the U.S., inspecting the wreckage, conducting tests, and reviewing thousands of documents. The testimony of the witnesses clearly brings to light the events that led to the failure; the survivors describe in their own words what they experienced during the harrowing fifteen seconds it took for the structure to collapse. Of the eighty-six men working on the bridge, only eleven survived—one of the survivors died later from his injuries. It was one of the worst industrial accidents in Canadian history.

The findings of the Royal Commission are discussed. The three engineers attributed the failure to the defective design of critical components of the structure and blamed the contractor's design engineer and the owner's consulting engineer for the failure. But the circumstances surrounding the preparation of the specifications, the design, the construction, and, most of all, who had final authority on the project, point to much broader culpability on the part of the organizations involved, including the government, and the individual, inexperienced decision-makers.

CHAPTER 1

PROFESSOR HAULTAIN'S REQUEST

THE STORY OF THE RITUAL of the Calling of an Engineer, including the Obligation and its symbol, the Iron Ring, begins with a remarkable Canadian mining engineer, inventor, and professor. Herbert Edward Terrick Haultain was born in England in 1869 and emigrated to Canada with his family in 1875. He studied at the School of Practical Science at the University of Toronto and helped organize the first Canadian Student Engineering Society, serving as its president from 1888 to 1889. He graduated with a degree in civil engineering in 1889 and did post-graduate work in England as well as in Freiberg, Germany, at the oldest mining school in the world. During his twenty years of active practice, he worked in Saxony, British Columbia, South Africa, Idaho, and Ontario. Early in his career, he worked at a tin mine in Bohemia where he designed and operated the first electric mining hoist ever used in continental Europe. While managing the Canadian Corundum Works in Ontario, he became interested in the problem and challenges of fine-sizing in the mineral industry. He invented two instruments to separate the individual minerals that make up an ore sample: the Superpanner, which separates very finely divided minerals according to their specific gravity, and the Infrasizer, used to separate extremely fine pulverized ore according to particle size. These instruments are still in use today.[1]

In 1908, Haultain was appointed professor of mining engineering at the University of Toronto, where he taught for over thirty years. Referred to affectionately by his students as the "Old Man," he had a reputation for being a great mentor and an inspiration to his students. He was notorious for his messy desk and he once remarked, "An empty desk denotes barrenness of soul." Haultain retired as professor emeritus in 1938 in order to devote all of his time to research. He died in Toronto on September 19, 1961, and is buried in Little Lake Cemetery in Peterborough, Ontario.[2]

Professor Haultain was a one-man employment agency for his graduating engineers. In the 1920s, almost a third of Canada's graduating engineers were emigrating to the U.S. In 1927 Professor Haultain and a work colleague, Robert A. Bryce, a noted mining engineer and president of Macassa Mines, co-founded the Technical Service Council, a non-profit, industry-sponsored organization whose aim was to find jobs for engineering graduates in Canada. This admirable organization placed hundreds of Canadian engineers and continued operating into the 1990s.

Haultain's influence over the mining industry cannot be overstated. He had a major impact through his own work and inventions, but perhaps more significantly through his more than three hundred mining graduates who became the engineers, managers, presidents, and pioneers of the mining industry in Canada. In 1994 he was recognized posthumously for his contributions to the mining industry when the Canadian Institute of Mining and Metallurgy inducted him into the Canadian Mining Hall of Fame. To recognize his contributions as an educator, a building is named in his honor at the University of Toronto.[3]

Professor Haultain's contributions to the engineering profession were significant and came at a critical point in the history of the profession. He belonged to the Institution of Civil Engineers of Great Britain and in 1925 was elected president of Ontario's licensing body, the Association of Professional Engineers of Ontario. He was also very active in the Canadian Society of Civil Engineers (CSCE), the first engineering organization in Canada. The CSCE was created by a group of engineers in 1887 with a mandate to "uphold the honor and dignity of the profession." CSCE's by-laws made it clear that "civil" included all types of engineering, except military. The organization's headquarters were at McGill University in Montreal. If you lived within fifty miles of the headquarters, you were considered a resident and paid an annual fee of eight dollars. If you were outside that fifty-mile radius, you were a non-resident and you paid six dollars. Students paid one dollar. CSCE branches spread across the country and membership reached 2,750 by 1910. Fortunately for the CSCE, Professor Haultain chaired its most important panel, the Committee on Society Affairs.

PROFESSOR HAULTAIN'S REQUEST

Mid-way through the First World War, CSCE's members felt that the organization was not meeting the needs of a growing number of its younger engineers, especially those in the burgeoning mechanical and electrical disciplines. The CSCE board asked its Committee on Society Affairs to examine the issues and report back with recommendations. Professor Haultain's committee did so in 1917, and proposed a number of significant changes, including a name change to reflect the increasing diversity and interests of the membership. It further proposed the publication of a monthly magazine featuring technical articles from its members, and most importantly, the appointment of a full-time secretary to keep things moving. It was this committee's diligence that led to the transformation of the CSCE into the Engineering Institute of Canada (EIC) in 1918. The revised mandate of the EIC was to increase the professional standing of engineers through public service, greater publicity, and, most importantly, through licensing.

On January 25, 1922, Professor Haultain was the luncheon speaker at the annual meeting of the EIC at the University Club in Montreal. The title of his talk was "The Romance of Engineering." This luncheon was the only event that women were invited to. In his remarks, he noted that this was the thirty-fifth year that the Annual Meeting of the EIC, formerly the CSCE, was being held at the "headquarters" in Montreal. Haultain described EIC's headquarters as being an autocracy of leadership rather than domination, of influence rather than by virtue of jurisdiction. Having complimented his hosts, he turned his attention to the ladies in the room. He made it clear that he was not, and had never been, in favour of opening the EIC meetings to the ladies, but then he declared, "I have seen a new phase of womanhood!" He heaped praise on women for their extraordinary efforts and the successes they had achieved in forcing their way into places and organizations where they were "very much less welcome than you have been here." He then told his audience about the "Girls in Green."[4]

Canada suffered more casualties in World War I than it did in World War II. The magnitude, brutality, and unbelievably harsh conditions of WWI are likely unparalleled in all of history. Many of Canada's war veterans returned home with severe physical and mental wounds, and they were left to recover in hospitals and convalescent wards. When pronounced physically and mentally well enough to leave the hospital, they were far from able to look after themselves or to lead productive lives. Occupational therapy was a relatively new concept at that time and had only been used to a limited extent in hospitals, such as the College Street Military Hospital in Toronto. In 1917 and 1918, it became clear that thousands of soldiers returning from the war would benefit from such therapy. This meant that many more workers trained in this field were needed. The engineering faculty of the University of Toronto, and Professor Haultain in particular, acting entirely without precedent, formed a committee to address the shortage of occupational therapists. Backed by an Order in Council and provincial government funding, classes in occupational therapy began in the spring of 1918. After eighteen months, 375 "Girls in Green" had successfully completed the eight-week program. The "Girls" were paid sixty-five dollars a month while in training, and eighty-five to a hundred and twenty-five dollars a month when they were instructing. The medical officer at the psychiatric hospital in Newmarket reported that six of these "Ward Aides," as they were referred to by hospital staff, in one month's time had increased the prospects of cure by fifty percent. *Maclean's* magazine published an article on February 15, 1922, titled "God Bless the Girls in Green," praising those responsible and singling out one individual: "Upon Professor H.E.T. Haultain, Vocational Officer for Ontario, fell the responsibility for the whole plan, and his enthusiasm, energy, and boundless optimism piloted the scheme to success."[5]

The *Maclean's* article appeared less than a month after Haultain's speech to the EIC, and so the subject was undoubtedly foremost in his mind. He told his luncheon audience that, in his view, these women saved men's souls, "not from the hell-fire of the future but the hell-fires of today." They helped these soldiers become self-sufficient and productive members of society. He noted, "To most of the girls, this work and its results was true romance of a very deep kind, in many cases, I firmly believe, a truer and richer romance than the personal affairs of the heart." But, Professor Haultain asked his audience, what did this have to do with the EIC and its meeting that day? He told them that there was no similar spirit among engineers; he appealed to the ladies in the audience by stating that "With a proper development of the tribal soul we can become anything in the community. Without it, we are very little indeed; the majority of us are hewers of wood and drawers of water—some possibly nothing but sounding brass. Now there is your romance. Help us find our tribal soul." No doubt encouraged by the successes achieved by the Girls in Green and their impact on soldiers returning from the war, he considered the role that women could play and closed with the following suggestion: "Just one word more, not a word of advice, simply a suggestion, or rather a statement of fact. There are in this city about two hundred young men graduating each year from the university into the profession at the most susceptible period in their lives, not conscious that there is such a thing as an engineering soul. Women could do anything with them." He ended his remarks with a plea to the ladies to help the engineering profession find its tribal soul.

Later that evening, Professor Haultain attended the EIC retiring president's dinner at the University Club. Seated at the head table were seven past-presidents of the EIC. The chairman was the immediate past-president, J.M.R. Fairbairn, chief engineer of the Canadian Pacific Railway. He had invited his friend Haultain to the dinner and asked him to elaborate on his theme of a "tribal spirit." Haultain suggested to the small group that the profession should develop an oath or creed to which the young graduating engineer could subscribe, something akin to the Hippocratic Oath, to welcome the young graduates into the profession. Haultain's two brothers were medical doctors and through them he had learned something of the meaning and importance of an oath or creed to bind together the members of their profession.

Professor Haultain challenged the seven to strike a committee to act on his proposal and to draw up some words that the young engineering graduate could adopt and learn by heart—something in the form of an oath or creed, or part of a ritual, representing his becoming a member of the tribe. Haultain made a motion, which was seconded by Brigadier-General C.H. Mitchell, and put to the group by Fairbairn. It was accepted as a formal motion and carried unanimously. The committee of the seven past-presidents was thereby struck on January 25, 1922. Fairbairn was named as chairman of the committee and the other six were R.A. Ross, H.H. Vaughan, G.A. Mountain, G.H. Duggan, Phelps Johnson, and W.F. Tye.

Having previously arranged with the hall porter to have a sleigh waiting for him outside the club, Professor Haultain then ran to catch his train to Toronto.[6]

Life and busy work schedules prevented the committee from meeting to develop the idea. In fact, nothing happened until October of the following year when Professor Haultain wrote to Chairman Fairbairn and asked him if anything had been done towards the development of a creed or professional oath for the young engineer. In his letter, dated October 4, 1923, Professor Haultain wondered "if it would be possible to interest Kipling in this."

Fairbairn responded immediately, expressing great enthusiasm for the idea. He wrote, "Why not write to him on the subject? I do not know anyone who would put it up in better form. He would certainly be doing a great thing for a class of professional men, of whom we believe he has a keener appreciation than any other prominent man of equal capacity to do the thing of which we are thinking."[7]

In his letter to Rudyard Kipling, dated October 19th, 1923, Professor Haultain described the task that the seven past-presidents had set for themselves, "to draw up some form of words that the young graduate in engineering could accept and learn by heart—something in the form of an oath or creed or part of a ritual, representing his becoming a member of the tribe." He attached Fairbairn's letter, which explained that the committee had been unable to get together to discuss the matter. Professor Haultain closed with the following entreaty: "We are a tribe—a very important tribe within the community, but we are lacking in tribal spirit, or perhaps I should say, in manifestation of a tribal spirit. Also, we are inarticulate. Can you help us?"[8]

CHAPTER 2

RUDYARD KIPLING'S RESPONSE

IN THE FALL OF 1923, Rudyard Kipling was fifty-seven years old and was, by all accounts, the most popular English writer in the world, and one of the most prolific. He had won the Nobel Prize for Literature in 1907 and was a close friend of the king of England, George V, and of the late American president, Teddy Roosevelt. Kipling had been offered a knighthood on several occasions, as well as the British Poet Laureateship, all of which he had graciously declined, preferring the simple life. Haultain and Fairbairn were knocking on the door of the greatest author alive, and perhaps the humblest. He was, to say the least, a fascinating man, and an acknowledged literary genius.

Professor Haultain's letter arrived at Bateman's, the Kiplings' home in Sussex, England, near the end of October. Kipling had just completed his two-volume, *The Irish Guards in the Great War,* a task that had taken him two years to complete.[1] One can only imagine what must have gone through the author's mind when he read Haultain's letter. Here was a representative of Canada's engineering profession asking for help to develop an oath or creed to bind engineers. Kipling, renowned for his love of engineering, machines, and all things tribal, but most of all for his belief in one's duty to one's calling as expressed in so many of his great works, was being asked to help develop a creed for engineers. Fairbairn had been right- there was no one in the world better suited to the task.

Joseph Rudyard Kipling was born to English parents in Bombay, India, on December 30, 1865. His name came from the place where his parents met, Lake Rudyard in Staffordshire, England. He spent the first six happy years of his life with his family in India, where his father taught, and was principal at an art school in Bombay. As was the custom for British expats, the day came when Rudyard and his younger sister were sent off to England for their education. Unfortunately, the boarding house where they lived became a prison for the young boy who was anything but favored by his keeper, Mrs. Holloway—or, as Rudyard called her, "the Woman!" But this would become the place where a writer was born. His punishment for various and sundry infractions, such as "showing off," was to be sent to his room to read the Bible and the Collects. After a time, it occurred to Rudyard that reading was a means to everything that would make him happy, and so he read all that came within his reach. Instead of dampening the boy's spirit, reading ignited his mind, and out of that painful experience, Rudyard Kipling, the Writer, was born. And what a writer! He entertained and enlightened young and old with his stories, starting from the time he was a teenager working at a newspaper in India until his death in England in 1936.[2]

All of Kipling's two hundred and fifty short stories carried a moral or lesson, as did his countless poems and several novels, including his timeless works, *The Jungle Books*, *Kim*, and the poem "If." He endowed animals, ships, and machines with human qualities and faults, and delivered his messages in an imaginative and unforgettable way, leaving an indelible impression on the reader.

At the age of twenty, Kipling was admitted as a Freemason; this association is key to understanding the man as well as his response to Haultain's request. In caste-ridden India, Freemasonry was the only ground on which adherents of different religions could meet "on the level." Kipling described his experience: "I was entered by a member of the Hindu, passed by a Mohammedan, and raised by an Englishman. Our Tyler was an Indian Jew." The idea of a secret bond, a sense of community and of high moral principles among men sworn to a common purpose, fit Kipling's concept of a social order.[3]

Kipling's lively imagination coupled with his deep sense of responsibility and strong work ethic made him the perfect choice. Very early in his career, he had become the voice of the new generation of engineers, bridge-builders, road-makers, miners, prospectors and ranchers who were doing their duty to advance the cause of civilization. He would seek out these doers wherever he went and, with his inquisitive mind, draw out the very essence of their craft and skill. He then used this knowledge to write his stories.

In "The Bridge Builders," Kipling tells the story of a Scottish engineer who designed and built a bridge across the Ganges. In it, the author displays his knowledge of engineering and his deep respect for the engineer and his work ethic. In 1894 he wrote "McAndrew's Hymn," a poem about a marine engineer hard at work tending to his ship's engines, in which the old engineer reminisces about his life while his "purrin' dynamos" beat out their message: "Law, Order, Duty an' Restraint, Obedience, Discipline!" For Kipling, the hero on the ship was the engineer down below, not the gold-laced captain on the bridge.[4]

Above all, Kipling was simple and humble, traits that he fostered in himself and admired in others. On June 22, 1897, Britain's Queen Victoria was celebrating sixty years as monarch of the British Empire. Kipling wrote a poem for the Diamond Jubilee titled "The Recessional." In typical Kipling fashion, he reminded his countrymen that "humility, not pride, awe not arrogance, and a sense of transience not a sense of permanence were to be the keynotes of the imperial festival." The poem made headlines, and it made him the de facto poet laureate of the British Empire.[5]

The great author had a particular fondness for Canada, crossing the country on each of his several visits. In 1907, he travelled to Montreal with his wife to receive an honorary degree from McGill University. At convocation he gave a speech titled "Values in Life," in which he admonished his audience against the acquisition of wealth for wealth's sake. He also spoke of the young person's soul, counseling them to "Take anything and everything seriously, except yourselves."[6]

The couple then headed west in a private Pullman car, as guests of Sir William Van Horne, the head of the Canadian Pacific Railway. They stopped in Winnipeg where Kipling gave another speech to the Canadian Club titled "Growth and Responsibility." He told his audience, "Your own labour, your own sacrifice, have given you material prosperity in overwhelming abundance. There is no man, and here I must quote again, that can foresee or set limits to your destiny. But any man, even I, have the right to remind you that to whom much has been given, from them much—much—shall be required."[7]

Soon after his return home from Canada, Kipling received his letter from the Swedish Academy offering him the Nobel Prize for Literature, awarded for the first time to an English writer. Kipling accepted—he was just forty-one years old, the youngest recipient of the Nobel prize to date. The Academy's citation reads,

"To the greatest genius in the realm of narrative that England has produced in our times. In consideration of the power of observation, originality of imagination, virility of ideas and remarkable talent for narration which characterize the creations of this world-famous author."[8]

That same year, Kipling penned one of his best-known poems, "The Sons of Martha." In it, he comments on the biblical story of a visit by Jesus to the home of Martha and Mary. While Martha busied herself preparing a meal for her guests, her sister Mary sat at the foot of Jesus, listening to him preach. Martha complained and asked Jesus to tell Mary to help her. Jesus replied that Mary had chosen the good part, which must not be taken away from her. Martha's duty was work. For Kipling, engineers are the Sons of Martha, among those who bear the burden of the world's work, making our lives more livable by providing for our physical necessities.[9]

Kipling and Lord Robert Baden-Powell, the founder of the scouting movement, were close friends. In 1914, Baden-Powell announced the establishment of a junior section for scouting, the Wolf Cubs. He asked Kipling for the use of his *Jungle Book* history and universe as a motivational frame for cub-scouting. Kipling was delighted.[10]

During his career, Kipling travelled extensively and came into contact with most, if not all, of the leading English and American authors of his day, developing lifelong friendships with many of them, including Henry James and Thomas Hardy. During his first trip around the world, he crossed the U.S. and showed up at Mark Twain's doorstep, not knowing what reception he'd get. Twain liked Kipling and later wrote about their meeting: "Between us, we cover all knowledge; he covers all that can be known, and I cover the rest." He also gave the budding author some advice: "Get your facts first and then you can distort 'em as much as you please."[11]

Kipling's daughter, Elsie Bambridge, gave her permission to Charles Carrington to write a biography of her father. Among the biographer's many observations, this one stands out for those who wear the symbol of the Obligation: "The strongest continuing motive of Kipling's work throughout his whole career was the sense of comradeship among men who share a common allegiance because they are committed to a common duty. His love of technical jargon, though partly a mere delight in the richness of language, was strengthened by a conviction that the best talk is 'talking shop,' the kind of conversation which lives and is genuine because it is based upon a secret shared between those who know how something is done, a secret conveying that kind of knowledge, which is power."[12]

KIPLING'S RESPONSE

Professor Haultain's letter was dated October 19, 1923. It would have taken five to ten days for it to reach England. Kipling's response was dated November 9th. Haultain had been absent from his office for two weeks, and when he returned on November 27th, the package from Kipling was sitting on his desk. He opened it and couldn't believe what he saw—it was beyond anything he had ever imagined or hoped for. It was titled "𝕿𝖍𝖊 𝕽𝖎𝖙𝖚𝖆𝖑 𝖔𝖋 𝖙𝖍𝖊 𝕮𝖆𝖑𝖑𝖎𝖓𝖌 𝖔𝖋 𝖆𝖓 𝕰𝖓𝖌𝖎𝖓𝖊𝖊𝖗." The text had been hand-written in Old English lettering by the great author himself. Haultain was speechless. He sat down and read every word, in awe of what the great Rudyard Kipling had accomplished. The package included the Obligation, the text and a description of the Ceremony, a detailed description of the Iron Ring and how it was to be worn, together with notes about the Ritual. Kipling preferred the word "Obligation" to the word "Oath" or "Creed." The renowned author had taken less than two weeks to come up with the entire package. Haultain and Fairbairn had chosen well, indeed.[13]

Haultain sent Kipling a note: "The first impression is that it is very different from anything we could have thought out for ourselves and that it will take a little time to appreciate its full significance and beauty. This is as it should be." He then made a copy for himself and immediately sent the author's original work to Fairbairn in Montreal.[14]

Fairbairn was astonished at how quickly the response had come, but most of all, he was deeply impressed with the work done by none other than Rudyard Kipling. He sent copies to his six committee members, who were all travelling abroad. Eventually, each of them signaled their heartfelt approval, but try as he might, Fairbairn was unable to get them together to plan the next steps. The package lay dormant for almost a year.

The following October, Professor Haultain once again took the initiative and sent a letter to Fairbairn, asking for a meeting to discuss the package from Kipling, with the intention of having the first Ritual ceremony take place the following spring at the University of Toronto, following convocation. Fairbairn promised Haultain he would meet with him, but first he wanted someone to draft a procedure, or rules to be followed, at the first Ceremony. In November, one of Fairbairn's committee members, Robert Alexander Ross, came up with a set of guidelines. However, Fairbairn and Ross were reluctant to proceed without a meeting and full discussion with all seven members of the committee. Although he and Ross were in Montreal, the other five were in Italy, France, South America, California, and one was in hospital. Finally, Haultain wrote to Fairbairn on March 23rd, urging him to act in spite of this and pointing out that he had their tacit approval to proceed; but Fairbairn and Ross resisted.[15]

On March 26th, no doubt encouraged by Haultain, the Council of the Engineering Alumni Association of the University of Toronto sent a telegram to Fairbairn advising him that "The Council heartily approve and desire immediate action to enable them to use the Ritual with this year's graduating class in engineering on May 1st. Will you take necessary action to authorize us?" The pressure was mounting. On that same day, Haultain offered to travel to Montreal to meet with Fairbairn and Ross. Fortunately, they agreed.[16]

On the morning of March 31st, Haultain and two engineers from the Alumni Association sat in Fairbairn's office in Montreal, with Ross on the phone. The group meeting began at 9:15 a.m., and for the next thirteen hours, they studied Kipling's package, reviewing and discussing at length the procedure to be followed and how the text should be read during the Ceremony. A copy was made of the Ritual, upon which they made some slight changes in red. By 10:30 p.m. that night, they had a package ready to be sent to Kipling, as well as a plan to proceed with the inaugural ceremonies that spring. Haultain and his colleagues had managed to convince Ross and Fairbairn that they had sufficient authority to move forward.

Fairbairn's cover letter to Kipling explained the reasons for the few changes proposed and asked for his approval. He also asked for his approval of the draft statement of procedure; nothing was going ahead without Kipling's say so. Fairbairn ended his letter "assuring you of our deepest appreciation of your great kindness in this whole matter and of our intention to make it go with the maximum of dignity."[17]

Kipling was travelling in France at the time. The cable went to his address in Paris just as he was leaving, and it then followed him home to England. True to form, as soon as he received and reviewed the package, Kipling cabled Fairbairn on April 22nd: "Amended Ritual and Statement of Procedure approved. Regret delay answering owing to absence abroad." He also wrote to Fairbairn the same day, confirming his approval and apologizing again for his delay in responding.

Immediately upon receiving the wire message from Kipling, Fairbairn sent a telegram to Haultain: "We're off. Kipling's approval received this morning."

The stage was now set for the inaugural ceremonies, the first of which was to take place in Montreal in just three days.[18]

Ross and Fairbairn went to work immediately. Ross memorized the entire Ritual while Fairbairn contacted four of his classmates from the University of Toronto's class of '93 and gathered together the necessary accessories to be used during the Ritual. On the evening of Saturday, April 25th, Ross, Fairbairn and his classmates had a quiet dinner in a private dining room at the University Club in Montreal. Afterwards, in a Ceremony supervised by Ross, the six engineers took the Obligation concurrently.[19]

After getting the go-ahead from Fairbairn, Haultain and his colleagues sprang into action as well; they gathered together the necessary articles required for the Toronto Obligation Ceremonies and made all of the arrangements. On Friday morning, May 1st, three of the newly Obligated engineers from Montreal travelled to Toronto to conduct the ceremonies. At 11:15 a.m., in the Senate Chamber of the University of Toronto, fourteen officers of the University of Toronto Engineering Alumni Association were Obligated, with Fairbairn and Ross presiding. Immediately after receiving their diplomas that afternoon, 107 new engineering graduates proceeded to the University Senate Chamber where they took the Obligation in a Ceremony led by Ross. Four prominent engineers from the Toronto area who had missed the morning ceremony took their Obligation and received their Rings with the new graduates. The Rings were placed on the candidates' fingers by Fairbairn.

On May 20th, Fairbairn sent a complete report to Kipling. In closing, he wrote, "I only wish that I could adequately express my own deep appreciation of what you, in giving us this Ritual, have done for us as a profession." Haultain had also sent a favorable report about the ceremonies to Kipling on May 12th. He explained the reasons for the delay between the time that he had received Kipling's package in November 1923 and Fairbairn's letter to Kipling in March 1925. There were two main reasons for this, he wrote: The first was "the natural inertia in a matter of this kind of such busy men as Fairbairn and Ross;" next was "the fear of making a slip… To us all your action was very precious. Fairbairn and Ross have a deference and a reverence for you and your work which amount to real worship and I am very sure that in the case of Ross this seriously delayed action. There was also the difficulty of getting the Seven together which grew worse with time."[20]

Haultain enclosed two samples of the Iron Ring in his letter to Kipling. He explained that they had been produced and hammered in the occupational workshops of the Christie Street Military Hospital in Toronto by returned World War I veterans who had suffered crippling injuries. These men had been trained for this work by the "Girls in Green" who had received their training from engineers at the University of Toronto. Referring to the roughness of the Rings, Haultain wrote, "No doubt the next batch of rings can be turned out with a better finish."[21]

In his reply to Haultain, Kipling wrote, "The rougher the Ring the better. It is not half so rough as the life ahead of most of the boys." In early June, Kipling also responded to Fairbairn's reporting letter with advice regarding the engineers administering the Ritual, as well as the Ring. He referred to the Ring samples he had received from Haultain, and his remark about their being rough. Kipling wrote, "The Ring is an allegory in itself. It is rough, as the mind of the young man. It is not smoothed off at the edges, any more than the character of the young. It is hand hammered all round—and the young have all their hammerings coming to them. It is his neither beginning nor end, any more than the work of an Engineer, or, as we know, Space itself. I would not depart by a shade from the make and nature of the Original Ring." He closed with, "Let us pray that it (the Ritual) will be a real success and a true aid to the calling and work of every Obligated Engineer."[22]

Kipling's use of cold iron as a symbolic metal for the articles used in the Ritual, as well as the Ring, stemmed from his interest in iron as a metal of power and a symbol of human innovation. There is an ancient belief that there is something magical about metals, and iron in particular. Kipling tapped into this tradition in "The Knife and the Naked Chalk," a short story set in the Weald forest where the earliest English iron had been smelted. Bateman's, the estate where the Kiplings lived, had been built by an ironmaster in the early seventeenth century, and was on the edge of the Weald forest. Traditionally, iron was a means of warding off the supernatural. It was believed that spirits hated cold iron, for if they touched or crossed over or under it, they would lose their magical powers. People in houses feared the spirits and warded them off by nailing horseshoes over their doors. "Cold Iron" is a short story written by Kipling in 1906 in which a young boy loses his magical powers when he finds a ring made of iron and clasps it around his neck. He must henceforward go among humankind and alleviate their sorrow as best he can. In the poem of the same name, Kipling makes a comparison between the boy's loss of his magical powers and Christ's human incarnation.[23]

In the months following the first Obligation ceremonies, Ross and Fairbairn drafted the Rule of Governance, as suggested by Kipling, to ensure that the Ritual would be administered consistently throughout Canada. In October, Haultain wrote to Kipling advising him that Fairbairn planned to travel to England before the end of the year, and that he hoped to meet with Kipling to go over the Rule of Governance. Kipling replied that he looked forward to seeing Fairbairn in England. Near the end of 1925, Fairbairn spent a day with the Kiplings at Bateman's. He and the author carefully reviewed and made changes to the draft Rule of Governance. Soon after his return to Montreal, Fairbairn worked with Ross to complete a final draft with Kipling's suggested changes. Following approval from the other five Wardens, the draft was sent to Kipling on March 2, 1926. Kipling returned the papers soon after, "initialed para by para, so that there may be no doubt that I have gone through them and approved."[24]

In June 1930, the Kiplings were in Montreal, on a return trip to England from Bermuda. Some thought had been given to organizing an Obligation Ceremony that could be attended by the author, but at his request, the idea was abandoned. A private dinner, however, was organized by Fairbairn and Ross and attended by two other Wardens; Professor Haultain was not present. The author expressed great satisfaction with the status and acceptance of the Ritual in Canada.

"The Ritual of the Calling of an Engineer" is uniquely Canadian and was copyrighted in Canada on June 5, 1926, as a literary, unpublished work by Rudyard Kipling. The seven past presidents are listed as the proprietors. It was copyrighted in the United States in 1941. By letters patent issued by the Province of Quebec, the committee of the seven past-presidents of the EIC was incorporated on March 18, 1938 and is called "The Corporation of the Seven Wardens." The Iron Ring was copyrighted in Ottawa on June 29, 1949. Trademarks for the Ring's design were registered in Canada in 1961 and in the U.S. in 1965.

The Corporation of the Seven Wardens reminds all Obligated engineers that "the business of the Ritual, while no mystery, is neither for the public nor the press."

The first women to be Obligated took part in a Ceremony held at the University of Toronto in April 1927. By 1934, there were nine camps spread throughout

Canada; today there are twenty-seven. In 1952, Haultain wrote, "Twenty-six years after the first Ceremony more than twenty-five thousand had subscribed to the solemn Obligation, which is the central feature of the Ceremony. With the main features, the seed from Kipling's amazing genius, and the fertile soil of the Canadian Engineering spirit, the result followed logically."[25]

To date, more than half a million engineering graduates have been Obligated, on Cold Iron.

CHAPTER 3

A BRIDGE AT QUEBEC

"**KEBEC" IS HOW THE NATIVE** Algonquin people referred to the area where Quebec City is located, "where the river narrows." The St. Lawrence travels 750 miles from the foot of Lake Ontario to the Gulf of St. Lawrence, carving its way between the Canadian Shield to the north and the Appalachian and Adirondack Mountains to the south. At Quebec, the river narrows to a gap of just three-quarters of a mile and reaches a depth of over 200 feet. The banks of the St. Lawrence near the old city rise cliff-like 150 feet above the water. The St. Lawrence rises and falls as much as eighteen feet with the rhythm of the Atlantic tides, and winter ice jams can reach heights of fifty feet. Prior to 1918, the only way to get across the river between Lévis and Quebec City was by ferry or over a winter ice-road, but there were several weeks at either end of the season when even this wasn't possible.

The mid-to-late nineteenth century was a time when railroads united countries and propelled economic development. Lévis, on the south bank of the river opposite Quebec, gained access to the Grand Trunk Railway system in the east in 1855. A steam ferry made the connection between Lévis and Quebec City. Quebec, however, was still without a railway link to the west. It was a dead end and it would be another ten years before that gap was filled. The island of Montreal, 160 miles to the south, was far better placed to receive railway traffic. When Montreal built the Victoria Bridge across the St. Lawrence in 1859, giving the Grand Trunk

Railway a continuous route to Portland, Maine, and access to western ports, Montreal became Canada's leading eastern port; Quebec City, on the other hand, felt even more like the poor sister.[1]

In 1851, a determined Quebec City Council hired the well-known young engineer Edward Serrell to examine the St. Lawrence near the City and recommend a bridge site. Serrell had designed the Lewiston-Queenston suspension bridge that spanned the Niagara Gorge between Canada and the U.S. The site that was recommended by Serrell for the crossing was upstream from Quebec City. It turned out to be practically the same site that would be chosen many years later. Although it was seven miles upstream of Quebec City, it was considered ideal. This section of the river is less than 2,500 feet wide, with riverbanks 150 feet high, making it easier to achieve the navigational clearance required for ocean-going vessels. Serrell, of course, proposed to erect a suspension bridge for a single railway track and highway traffic. The span of 1,610 feet would be a world record for any type of bridge. Serrell viewed the undertaking as an absolute necessity. He concluded his report with, "Gentlemen of Quebec, you must either build a bridge, or a new city." But the economics weren't there and Serrell's three-million-dollar bridge proposal went nowhere.[2]

In 1867, with railway construction plans as a powerful incentive, John A. Macdonald was able to convince New Brunswick and Nova Scotia to join the Province of Canada, at the time made up of Ontario and Quebec. He took his Confederation proposal to the British Parliament, and the Dominion of Canada was created. By this time, a railway line was under construction on the north shore of the river between Montreal and Quebec, and more lines were in place on the south shore near Lévis. The need for a bridge between the two cities seemed obvious, but still beyond reach. Undaunted, Quebec City mobilized its politicians and business community and tirelessly lobbied Ottawa for a bridge. In 1884, a delegation representing Halifax, Saint John, and Quebec City traveled to Ottawa to press their case with the Privy Council and the prime minister.

Finally, three years later, in 1887, Macdonald's government created the Quebec Bridge Company by Act of Parliament. It was required to "build and operate a railway bridge across the St. Lawrence River" and to adapt it to the use of foot-passengers and vehicles. An act of parliament was necessary for two reasons: there was a government subsidy involved, and because the bridge would affect navigation on the St. Lawrence River. Construction was to start within three years and be completed within six; the site and all plans required the prior approval of the federal government.

Ten prominent and politically active businessmen joined the board of directors of the newly formed company and immediately turned their attention to the problem of financing the project. The act of incorporation required the company to obtain subscriptions totaling one million dollars of private stock. Shares were sold for $100 each, and within three months, the full million-dollar subscription was in hand; however, the actual money paid in by the shareholders never exceeded $65,000. Preliminary work on the bridge could now begin.

There was no one on the newly formed board with any relevant experience or technical knowledge. Edward A. Hoare, a British civil engineer with experience in railway construction, was hired to examine and recommend potential sites for the bridge. Hoare studied three sites upstream of Quebec City and submitted his report the following year. He recommended the same site Serrell had chosen in 1852, where the Chaudière River on the south shore drains into the St. Lawrence, seven miles upstream of Quebec City. The board submitted Hoare's report to Collingwood Schreiber, the chief engineer of the Federal Department of Railways and Canals.[3]

Schreiber endorsed the Chaudière site in his report to Parliament in February 1891. There were, however, many people in the local communities who were unhappy with the choice, preferring a bridge closer to Quebec and Lévis. This impasse resulted in further delays—the government refused to grant further subsidies until the proposed site was generally accepted. Unable to carry out the work within the allotted time, the Company would require an amendment to its enabling legislation to further extend the time for completion. This was the first of several amendments.

In the early 1890s, Canada went through a period of political instability: four prime ministers in five years. In 1896, Wilfrid Laurier, Member of Parliament for the federal riding of Quebec East, was elected Canada's first francophone prime minister. Laurier and his Liberals would govern Canada for the next fifteen years.

In 1896, with their bridge proposal stalled, two directors of the Quebec Bridge Company approached the mayor of Quebec City to join the board and to act as its president. Simon-Napoleon Parent had, according to one board member, "given so many proofs of his ability to do things, we might attain our end."

Parent was a well-known, well-respected, and well-connected lawyer, businessman, and municipal and provincial politician. He was born in Quebec City on September 12, 1855, into a family of modest means. Having received his basic education at a local school, he settled in the village of Saint-Sauveur where he opened a grocery store. He completed his classical studies in his spare time and entered law school at Laval University in 1878, all the while carrying on his business. Called "Pol" by his classmates, he could, at times, be aggressive; no one stepped on Pol's toes and got away with it. He graduated first in his class in July 1881, receiving the Governor General's gold medal, and he was called to the Quebec bar in August. Although Parent had excelled academically, he was very shy and soft-spoken—impediments in the practice of law. He overcame this disadvantage by partnering with capable barristers; he prepared his cases meticulously and his partners argued them in court.[4]

When a fire destroyed a third of Saint-Sauveur in 1889, the residents voted in favor of annexation to Quebec City to ensure they would receive proper municipal services. Shortly after, the premier of Quebec made Saint-Sauveur an electoral riding. The community now needed an alderman as well as a member of the legislative assembly. Parent was elected to the Quebec City council in 1890 and put his name forward for the legislature of the province. At the time, the leader of the federal Liberal party, Wilfred Laurier, was the Member of Parliament for Quebec East, which included the riding of Saint-Sauveur. In the hopes of benefiting from Parent's solid electoral organization, Laurier backed him, and Parent easily won the nomination.

The tireless Parent sat on several committees at City Hall, but he sat quietly as a backbencher in the provincial legislative assembly. Lacking in eloquence where rhetoric reigned, Parent worked hard and put his electoral machine at Laurier's disposal. In March 1894, Quebec City elected fifteen businessmen out of a council of thirty to represent their interests. The city's merchants admired Parent's dynamism, civic spirit, and practical good sense. He was seen as the man who could carry out their plans for regional development—railway links were seen as key factors. On April 2, 1894, Parent was elected mayor of Quebec City.

Parent joined the board of the Quebec Bridge Company in April 1897. Just days before the annual general meeting of the company in September, the president of the board resigned, clearing the way for Parent to be elected. Prior to doing so, however, Parent traveled to Montreal to meet with his long-standing political ally,

Prime Minister Laurier, to discuss the prospects of federal support for the bridge project. Laurier told Parent that he would make the bridge his personal affair, but he couldn't promise that the subsidy would be voted on at the next session of Parliament. He did promise, however, that it would be voted on at the following session. Satisfied, Parent attended the first annual meeting of the shareholders and reported his conversation with the prime minister to the board. Parent was elected president of the Quebec Bridge Company on September 2, 1897. He was able to convince a fellow member of the Quebec City Council, Ulrich Barthe, to join him on the board and act as its secretary.[5]

Following the most recent amendment to its enabling legislation extending the date to 1902, the company only had five years to complete the project. Without increased funding, however, the project could only pursue the early stages of survey and design. Parent was determined to keep things moving. He persuaded the board to hire Hoare once again to survey the chosen bridge site. Hoare took soundings of the river bottom and borings on land, established the position of the bridge and piers, and drew up the general outline plans. These plans showed a minimum clear span of 1,600 feet between the main piers and a height clearance of 150 feet for vessels, based on criteria set by the Federal Department of Railways and Canals for this section of the river.

In early June 1897, Hoare wrote to David Reeves, president of the Phoenix Bridge Company in Phoenixville, Pennsylvania. Twenty years earlier, while Hoare was the chief engineer of the Quebec and Lake St. John Railway Company, Phoenix had constructed bridges for his railroad. Since its inception in 1864, the Phoenix Bridge Company had built hundreds of bridges in the U.S. and Canada, including nine long-span bridges. In his letter to Reeves, Hoare asked whether any engineer of the Phoenix Company planned to attend the annual convention of the American Society of Civil Engineers to be held in Quebec City on June 30th. If so, he asked that the Phoenix engineer call to see him, in connection with "a project for bridging the St. Lawrence River near Quebec."[6]

John Sterling Deans, chief engineer for Phoenix, attended the convention and met with Hoare and several directors of the Quebec Bridge Company. One of the directors entertained the entire convention at his home on the St. Lawrence River near the bridge site. During the trip by steamer to his home, he pointed out the proposed site for the bridge and the steps the Company was taking towards its construction. Theodore Cooper, a leading American bridge designer from New York, attended the convention and was among the guests who saw the proposed site for the bridge.

Hoare told Deans during their meeting in Quebec that if Phoenix was interested in the project, he would be glad to send him a profile of the crossing at the proposed site and other necessary information so that "you may, if you wish, be prepared to bid, if the project is carried out." Deans received a package from Hoare shortly after his return to Phoenixville.

The Phoenix Iron Company, parent of the Phoenix Bridge Company, was founded in 1783 and built its plant in the town of Phoenixville, thirty miles northwest of Philadelphia, Pennsylvania. Initially, the Company produced nails, rails, Civil War cannons, and rifles. Eventually, it became involved in the design and construction of iron and structural steel for standard railway bridges in America, Canada, Japan, and elsewhere. Its first president, Samuel Reeves, invented the Phoenix Column in 1862. This was a hollow, circular column made up of either four, six, or eight wrought-iron segments that were flanged and riveted together. Inspired by British Navy masts, the resulting column was much lighter and stronger than the solid cast-iron columns previously used. They allowed for the construction of tall buildings on narrow urban lots, helping to facilitate the creation of the skyscraper and high-stress-load-bearing bridges. The success of the Phoenix Column led to the formation of a construction subsidiary named Clarke, Reeves & Company, which eventually became the Phoenix Bridge Company. In 1884, Samuel's son David Reeves became president of both the Phoenix Bridge Company and its parent, the Phoenix Iron Company. Whereas the Phoenix Bridge Company was an engineering and contracting company, the Iron Company manufactured the steel for the Bridge Company's projects. The work for the Quebec Bridge was to be done under this arrangement. David's brother, William Reeves, was the general superintendent in the Iron Works shops.

Similar to other American bridge companies, the Reeves brothers had developed a system for producing bridges that enabled them to keep their clients' costs down—production was key. The company's engineering department comprised about thirty engineers and draftsmen. Relying primarily on industry standards, the engineers produced designs that were easy to manufacture and erect; there was usually very little innovation involved. The rate at which manufacturing took place in the Company's shops depended entirely upon the availability of shop drawings. Erection at site and payment for work done was contingent on the availability of materials and the ease with which it could be erected. Phoenix had a reputation for working cheap and being on time, traits that suited the Quebec Bridge Company.

When Deans received the package from Hoare, he wasted no time. On December 7th, just five months after their meeting in Quebec, Deans sent Hoare Phoenix's preliminary general plans for a 1,600-foot span cantilever bridge, complete with a drawing of the proposed structure. The following month, the Quebec Bridge Company applied to the Railway Committee of Laurier's Privy Council for approval of the plans and the proposed site for the bridge. The plans that were submitted with the application were dated January 13, 1898, and signed by Parent, Barthe, and Hoare; the plan for the steel superstructure was the Phoenix plan. By Order in Council, the government approved the site for the bridge and the positions of the piers and abutments, as shown on the plans. The bridge would have a clear span of 1,600 feet and a clearance of 150 feet above extreme high water. Two railway tracks, two electric car tracks, two roadways, and two pedestrian walkways were to be provided.[7]

The Order in Council was dated May 16, 1898, and made it clear that before work could begin, all details had to be approved by the chief engineer of the Department of Railways and Canals. They were also subject to the approval of the Governor in Council, upon the joint report of the Minister of Railways and Canals and the Minister of Public Works.

In July, Hoare was instructed by Parent to communicate with the Department so that suitable specifications could be prepared for a call for tenders. Hoare went to Ottawa to see the chief engineer of the Department of Railways and Canals, Collingwood Schreiber. Hoare was instructed by Schreiber to "Go into Douglas and go over the specification with him." Robert C. Douglas was the chief bridge engineer for the Department of Railways and Canals, and had been since 1893. The department's standard specifications were founded upon general specifications drafted by Douglas in 1896. Hoare walked into Douglas's office with a rough draft of the specifications and asked him to go over it with him.

There were parts of Hoare's proposed specifications that Douglas didn't agree with. Hoare told him it didn't make any difference since this specification wasn't for the construction of the work itself. It was only going to be used for the call for tenders; new specifications of a different kind would be prepared when the construction contract was let. Douglas reluctantly agreed and made no changes. Hoare's specifications were sent to Schreiber who approved them without question, since Douglas had approved them. Tenderers would submit bids based on specifications that would be changed at a later date, prior to construction.

Circulars inviting fixed-price tenders were issued by Secretary Barthe on September 6th, with a closing date set for January 1, 1899. The circular included a drawing showing a section of the river, the clearances required, and specifications for a cantilever bridge; the package did not include the Phoenix drawing. Tenderers were instructed that if they wished to propose a suspension bridge, they were to furnish complete specifications together with their bid. Fixed, lump-sum prices were to be submitted for both the substructure piers and the steel superstructure. The time for closing of tenders was subsequently extended to September 1, 1899.

By the Railway Subsidy Act of 1899, the Dominion Government granted a $1,000,000 subsidy to the Quebec Bridge Company. Although the Company wasn't yet in a financial position to enter into a contract for any of the work, the board was confident that more funds would be forthcoming from the federal government.

Recognizing that the construction of this bridge was unprecedented, and that Hoare did not have the technical qualifications for a bridge of this magnitude, the board discussed the need to appoint a consulting engineer. Six prominent engineers were considered. On February 23, 1899, Barthe wrote to Theodore Cooper in New York and asked whether he "was at liberty to take up the examination of their competitive plans." Cooper replied that he was.[8]

CHAPTER 4

CHIEF ENGINEER

THEODORE COOPER WAS A RENOWNED American bridge designer whose career spanned several decades. He was considered by his contemporaries as the leader in bridge design. Born in New York on January 13, 1839, he graduated as a civil engineer from the Rensselaer Institute (now Rensselaer Polytechnic) in New York in 1858, at the age of nineteen. After three years with a railway company, he enlisted in the Union Navy and served as an assistant engineer on the gunboat Chocura for the last three years of the American Civil War. He then moved on to a teaching post at the U.S. Naval Academy, followed by a tour of duty in the South Pacific. He left the navy in 1872 and went to work as an inspector for Captain James Eads, designer of the Mississippi River steel-arch bridge in St. Louis, which at that time was the most ambitious use of the cantilevered method of bridge erection ever attempted. Eads saw promise in the young engineer and put him in charge of erection. Cooper didn't disappoint the captain, once going without sleep for sixty-five hours during a crisis. At one point during the erection, Cooper wired Eads at midnight to warn him that the arch ribs in the structure were rupturing, a potentially disastrous condition. Eads immediately wired back instructions to Cooper and the bridge was saved.[1]

Following completion of the St. Louis Bridge in 1874, Cooper's career flourished. He succeeded Eads as chief engineer of the Bridge and Tunnel Company and was assistant engineer of the first elevated railroads in New York City. In 1879, Cooper resigned as superintendent of Andrew Carnegie's giant Keystone Bridge Company in Pittsburgh and set up his own consulting office in New York City. He published several important works on railroad and highway bridge design and consulted on rapid transit systems in the Eastern U.S. In 1894, Cooper set the standard for the safe loading of railway bridges—the Cooper Loading System is still in use today. He was recognized for this and other accomplishments with two prestigious awards from the American Society of Civil Engineers. Cooper was a proud, confident man fiercely devoted to his career. He expected much from himself as well as from others and was known for his direct and unsparing criticism.

On March 23rd, 1899, Parent, Hoare, and Barthe travelled to New York City to interview the eminent engineer. They briefly described their plans for the bridge and asked Cooper how long it would take him to review the tenders and what his terms were. Cooper told them it would take him three months and that his fee was $5,000. They also discussed his continued involvement as consulting engineer and asked him if the inspection of the work was included in his services. He told them his fee would be $7,500 a year once the work started and that it did not include inspection. Nothing was said about expenses. They parted without coming to an agreement, but Parent left Cooper with the impression that the plans would be sent to him for review. Following the meeting, letters were exchanged, and Cooper received the plans and tenders that had been submitted by the bidders.[2]

Cooper had developed standard specifications to be used by bridge manufacturers, but his experience as an actual designer was limited. The longest span that he had designed himself was the Sixth Street Bridge in Pittsburgh in 1892, with a clear span of only 440 feet. In Cooper's view, it was perfectly acceptable and in keeping with general practice in America to place the design of a structure in the hands of the engineering staff of contracting companies such as Phoenix. He believed that no consulting engineer such as he could afford to maintain a staff of such character and that no owner would listen to a fee that would cover such expenses. This was exactly what the Quebec Bridge Company wanted; the consulting fees would be kept to a minimum as the cost of the design work would be included in the contractor's price. This was also what the contractor preferred as this meant that the design could be prepared to streamline its manufacturing operation.

During the next three months, Cooper examined four proposals: one from the Dominion Bridge Company of Montreal and three from American firms—the

Keystone Company of Pittsburgh (Carnegie Steel), the Phoenix Bridge Company of Phoenixville, and the Union Bridge Company of New York. Although three of the bidders submitted cantilever designs, the Dominion Bridge cantilever plan was identical to the one submitted by Keystone—the two companies had agreed that if either of them got the job, they would share the work. Three firms submitted separate suspension bridge designs. Cooper studied the proposals, interviewed the designers of each plan, and prepared a detailed report.

Two of the three suspension bridges had spans of 1,800 feet, and the one from Dominion Bridge had a span of 2,000 feet. Cooper dismissed Dominion's suspension plan due to "the relatively high tender, and also from the incompleteness of the proposal, due to the qualification made in reference to the construction of the cables." Union Bridge's suspension design was also excluded from further consideration due to "the indefiniteness and incompleteness of the tender, and also because the plan was not in accordance with the specifications." That left only Phoenix's 1,800-foot suspension plan to consider for this type of bridge.

Cooper thought that the Phoenix suspension plan was acceptable from a technical viewpoint. One of the chief concerns with suspension bridges is a lack of stiffness, making them susceptible to high winds.[3] The Phoenix suspension plan had been developed by Gustav Lindenthal, a leading American bridge designer. Lindenthal's design had addressed the stiffness problem by trussing the cables, which addressed Cooper's concerns. However, the price for Phoenix's suspension plan was $600,000 higher than its cantilever span. Cooper rejected it as being too costly. Having rejected the three suspension spans, Cooper now had to decide between the two cantilever proposals.

There were only two companies now competing for the contract to design and build the cantilever structure: Keystone and Phoenix. Following some re-working of Keystone's plan by reducing the live load to comply with the specifications, Cooper evaluated the two bids for the superstructure. In his report he noted, "The greater depth of the Phoenix design and the curving of the top members of the cantilever arms give this plan a more pleasing effect than is produced by the lower depth and straight chords of the Keystone plan." The Phoenix plan also "appeared to be more economical, not only in weight of metal, but in saving four feet in the length of the piers." The Phoenix bid was $2,438,612. Keystone's bid was $23,500 higher, a difference of less than one percent. The tenders for the superstructure were based on lump-sum prices. Tenderers also provided unit prices for steel, but these were only for purposes of calculating progress payments. Keystone's unit price was, however, lower than the Phoenix unit price.[4]

The substructure was bid separate from the superstructure and consisted of four piers, two on each shore. The anchor piers were located higher up on the riverbank and the main piers, which support the majority of the load, were on the river's edge.

On June 23, 1899, Cooper submitted his report to the Company with his recommendation: "I hereby conclude and report that the cantilever superstructure plan of the Phoenix Bridge Company is the best and cheapest plan and proposal submitted to me for examination." He also recommended acceptance of the plan and proposals for the substructure piers made by the local company of William Davis & Sons of Cardinal, Ontario.

In a supplementary report of the same date, Cooper pointed out that his examination of the competitive plans was based entirely upon the specifications and data furnished by the Quebec Bridge Company. He thought that before construction should begin, a careful study should be made to see if "a better bridge could not be had and whether a change of span was not desirable." He also recommended that more information should be obtained regarding the riverbed.[5]

The board of the Quebec Bridge Company wasted no time. Within days of receiving Cooper's report, they met and resolved that "a copy of Mr. Cooper's report, with superstructure plan of the Phoenix Bridge company, and the Wm. Davis & Sons' substructure plan, be sent immediately to the Right Honourable Sir Wilfred Laurier." Parent was keeping the prime minister informed. None of the board members were qualified to review the documents and neither was Hoare.[6]

Following Cooper's favorable report on the Phoenix Bridge Company's proposal, the board summoned the Phoenix chief engineer, John Deans, to Quebec City. After their meeting, Parent wrote to Deans on August 23rd stating that the Company was ready to enter into a contract with Phoenix upon certain conditions, including a modification to the specifications and to the terms of payment. Parent explained that without the financial backing of the federal government, the Company was not in a position to award the contract to Phoenix based on their fixed-price tender. Phoenix was asked to accept their share of the subsidies from the government, or their equivalent, and the difference in bonds. Phoenix declined Parent's offer to accept the securities of the Quebec Bridge Company as payment for work. Deans made a valiant effort to place the securities with several American financial institutions, but his efforts were in vain. The banks told Phoenix there just wasn't sufficient traffic and revenue in sight to justify the investment.

Deans wrote to Parent and offered to undertake the work in whole or in part at the unit prices named in their tender, for "say one or two years." There would also be an understanding, he added, that the prices would be modified based on the current price of metal, to be fixed by agreement between the engineers of the companies on the date of the final order for each part of the bridge. The Company's lack of financing had now put Phoenix's fixed-price tender in jeopardy. Phoenix was prepared to proceed, but only to a limited extent, and then only on a unit price, or cost-plus basis.[7]

The project stalled, but Parent and the directors of the bridge company went to work. Their efforts paid off when in March 1900, the Province of Quebec granted a subsidy to the Company of $250,000, on the condition that the city of Quebec do the same. Quebec gave the Company a $300,000 subsidy, provided that the Company lay its terminus within the city's limits. The board met shortly after, with the intention of concluding the contract for the masonry piers and to address the outstanding issues with Phoenix. The six-member board split into two teams to deal with the contractors. On April 12th, Parent's contingent travelled to New York City to meet with Deans in Cooper's office. A deal was struck and an agreement in principle was signed, based on the payment terms that had been proposed by Deans in his letter the previous August. The agreement stipulated that the superstructure and steel anchorages were to be ordered within three years. This was no longer a lump-sum contract, however. Phoenix would be paid a unit price for every pound of steel erected, at market rates. The risk of cost overruns resulting from escalation and higher-than-anticipated quantities now rested with the Quebec Bridge Company, not Phoenix.

Aware that the Keystone unit price was lower than the Phoenix unit price, the board of the Quebec Bridge Company justified, or rationalized, its decision not to call for new tenders by pointing to Cooper's recommendation to accept the Phoenix proposal, as it was "the most economical and satisfactory in every respect."[8]

Just two days after their meeting in New York, Deans anxiously wrote to Parent asking him if the board had approved the April 12th agreement. Deans also wanted clarification about the respective powers of Cooper and Hoare. "We understand that in all engineering matters, we are to receive our instructions from Mr. Hoare, your engineer, and that he works under authority from your board. Further, we understand that all of our detailed plans of the structure, including sections, etc., must have the approval of Mr. Cooper, consulting engineer. Please advise if we are correct in this."[9]

The board approved the agreement with Phoenix on April 21st. The arrangements with M.P. Davis for the masonry piers were also finalized, but that contract wasn't signed until June. Parent wired Deans that the agreement had been approved and so Phoenix could now proceed with plans for the steel to be embedded in the anchorage piers. Regarding lines of authority, he added, "You can confer with Cooper and Hoare re plans." Deans sought no further clarification; Cooper was to approve the plans, but Hoare was to decide all "engineering matters." The issue as to whose decisions took precedence, Hoare's or Cooper's, was not addressed. It seemed clear, however, that Hoare would defer to Cooper on all but financial issues.

As requested by Cooper, Hoare had carried out a more detailed investigation of the conditions of the riverbed. He submitted his report to Cooper in early 1900. On May 5th, three weeks after the Company signed the agreement with Phoenix, Cooper submitted his report to the board. He recommended that the main span for the cantilever bridge be increased from 1,600 feet to 1,800 feet. He had discussed the feasibility of the increased span with Hoare, who was in agreement, as long as, "The additional expense wasn't too great." In his report, Cooper confirmed that the Department of Railways and Canals' legal requirements, as well as the conditions of the river channel, dictated that the span could not be less than 1,600 feet. He pointed out that keeping to this minimum span meant that the foundations for the main piers would have to be sunk to a depth of ninety feet, and that the piers themselves would stand in water up to forty feet deep and be subject to full river ice flows. Increasing the span by 200 feet would mean that the piers could be located farther up the riverbank, making construction easier and reducing the risk of erosion and damage caused by spring ice flows. He estimated the cost savings for the construction of the piers to be about $400,000, since building the piers in the dry would save a year on the schedule. But increasing the span meant that more and heavier steel would be needed for the superstructure, adding $450,000 to the cost. Cooper made no mention of the fact that the increased span would make the Quebec Bridge the longest span bridge in the world—ninety feet longer than the Firth of Forth cantilever bridge in Scotland.

The contract with Phoenix was for a 1,600-foot span. Although this was a significant change, what had been a fixed-price contract was now essentially a cost-plus contract; the increased cost for the steel would be absorbed by the Quebec Bridge Company.

A few days prior to submitting his report to the board, Cooper had sent a letter to Parent suggesting that he, Cooper, "…be instructed to make such modifications in the accepted competitive plan when adapted to the new lengths, as may tend to reduce the cost without reducing the carrying capacity or the stability of the new structure." Cooper wanted to change the government's specifications in order to minimize the added weight needed to accommodate the increased span.

Cooper's recommendation to increase the span was adopted by the board the same day they received his report. At that meeting, the board appointed Hoare as chief engineer and Cooper as consulting engineer to the Company. Cooper's terms and conditions were to be those set out in the board meeting of March 1899. The terms discussed by the board were then reduced to writing and a letter was sent to Cooper. However, the terms and conditions set out in the 1899 letter addressed only the retainer to review the plans and tender submissions, not for Cooper to act as consulting engineer. There was no language in the letter, or anywhere else, addressing what authority Cooper had or whose decisions had precedence: his or Hoare's. Although Hoare had the title of chief engineer, and Cooper that of consulting engineer, in practice, Cooper would come to be seen as the de facto chief engineer. Absent clear lines of authority, the contractor could decide who it would take direction from to suit its own purposes. In fact, no contract was ever drawn up for Cooper's services—there was a series of ten letters that made up the agreement between the parties, and these were never even bound together.

Hoare started his career as an engineer on marine works in London, England, in 1866, emigrating to Canada in 1869. Although his career spanned thirty-five years in railway works and included 27,000 feet of bridging, the majority of which was of steel on masonry piers and deep-water foundations similar to the Quebec Bridge, the longest bridge he'd been involved with was the Hawkesbury Bridge over the Ottawa River with seven spans of 210 feet each. The longest clear span he'd ever been involved with was a 300-foot bridge over the Lachute River. His involvement with these structures focused primarily on construction and management; he had little or no design experience. In spite of this, the board appointed him as chief engineer for what was to be the longest span bridge in the world. Hoare described what he saw as his role on the project: "to take general charge of the work; to undertake the duties generally undertaken by the engineer for a company; that is, to make surveys for the work, plans, specifications, the latter to a limited extent, prepare for contracts, see the work carried out, and to make progress estimates for payment to contractors." In fact, all matters having to do with plans and specifications were to be dealt with by Cooper. The majority of Hoare's time would be spent on preparing progress estimates and reports for the board of directors.

Within a week of receiving Cooper's report and recommendation, Phoenix accepted the increased span. Deans instructed his bridge designer to prepare a preliminary design for an 1,800-foot span, the longest span in the world. Phoenix agreed to deliver the steel anchorages to be embedded in the anchor piers within four months following the approval of the detailed plans. They also agreed to complete all general and detailed plans for the entire superstructure "with all possible speed." However, without further government subsidies, Phoenix was not prepared to expend the effort to carefully review and amend its plans for the increased span. Besides, Phoenix was busy with other projects, such as the Manhattan Bridge in New York.

Laurier's government had to grant the Company yet another extension in 1900, re-setting the time for completion to June 1905. At the Quebec Bridge Company's annual meeting in September, the board re-elected Simon-Napoleon Parent as president. On October 2nd, 1900, during a grandiose ceremony, Prime Minister Laurier, with Parent at his side, laid the "corner stone" for the north anchor pier of the Quebec Bridge. The *Globe and Mail* reported that the ceremonies were witnessed by an immense crowd, and that the mayor of Montreal congratulated Quebec upon acquiring its bridge and "ridiculed the idea that there should be any jealousy between Montreal and Quebec. The prosperity of one meant that of the other also."[10] The following day, Parent became the premier of the province of Quebec. He now wore three hats: premier of Quebec, mayor of Quebec City, and president and chairman of the board of directors of the Quebec Bridge Company. Less than a week after the corner stone was laid, Laurier's Liberal government granted the Company another subsidy of one million dollars, one-third of which was to be spent on the substructure and two-thirds on the superstructure.

Later that year, a second contract was entered into between the Quebec Bridge

Company and Phoenix for the erection of the approach spans on each side of the river. The unit price for the work was 4.114 cents per pound of steel, manufactured, erected, and painted.

More than a year after his appointment, the issue of Cooper's fee was still not settled. He wrote to Hoare in July 1901 and suggested that he be paid a lump sum amount of $22,500, or $7,500 per year, for the period from April 1900 to the completion of the metal superstructure, with an additional fee of $2,500 for each year exceeding three years that his services were required. Cooper's proposal made it clear that the Company could not call upon him to be out of New York City more than five days per month. In August 1901, Cooper was in Quebec City—his fees were overdue. He found the Company "embarrassed for funds." Considering that "it might be some years before any actual and important work would be required from me as consulting engineer, I wrote a new offer which amounted to reducing my fee to one-half. A member of the board suggested it be rounded out to $4,000." Cooper's proposal was accepted and approved by the board on August 7, 1901. The lower annual fee was never revised, and it would have to cover Cooper's time as well as that of his assistant, an expense that was never discussed.[11]

When Cooper presented his first invoice under the new arrangement to the Company, Secretary Barthe questioned the amount billed; he pointed out that Cooper hadn't been five days in Quebec. Cooper told Barthe that the clause had no such meaning; it was not to be interpreted to mean that he must spend five days in Quebec every month. He explained to Barthe that "Parties out of New York do not value the time of a consulting engineer as of any importance, and when called to a distant point for consultation on work for which I was acting as consulting engineer, I found great waste of time; the directors would not think it important to meet at the time stated, they would postpone the meeting for a week and think it my duty, being their consulting engineer, to await their convenience. Several times during my visits to Quebec I have left three or four days before a meeting of the board which was postponed, my good friends assuming that I would enjoy that spare time at Quebec, forgetting that I had other business of importance to devote my time to."

Cooper visited the site three times during the sinking of the caissons for the bridge piers. After the piers were completed, however, he never visited the site again. He was not present at site during the entire erection of the superstructure, citing illness as his reason for not travelling, "on doctor's orders." The illnesses most often mentioned were "grippe" and "fatigue." Cooper's health may have been fragile, but he was hardly an invalid, commuting almost daily by foot to his office at the foot of Manhattan Island from his home on West Fifty-Seventh Street.[12]

Over the next two years, work progressed on the substructure piers and anchorages. Cooper and Hoare gave favorable reports of the progress and in October 1902, Cooper reported that the experience of the last two years amply justified the change in length of span from 1,600 to 1,800 feet. During this time, Phoenix built the two smaller approach spans, which were simple spans supported at each end, one on land and the other on the anchor pier. While Cooper was busy preparing the specifications for the big spans, in order to save time and get things underway, the design for these smaller structures was based on specifications provided by Douglas, the Department of Railways and Canal's chief bridge engineer, and not on the original specifications issued with the tenders.

Negotiations for the main span of the superstructure had stalled. Phoenix refused to enter into further contracts until the Company had secured the necessary financial backing from the government to fund the entire project. But the government was as yet unwilling to provide the necessary guarantees to secure payment. Davis, the local contractor building the piers, was having difficulty getting paid for the limited work he was doing. He did some work in 1902 but stopped in the fall due to lack of money. He was even asked to carry some of the cost himself to keep things moving. Some of the money earmarked for Davis's work had been paid to Phoenix to keep them going on the approach spans.

By the end of 1902, things were at a standstill. Everyone looked to Ottawa.

CHAPTER 5

SPECIFICATIONS AND FINANCING

IN 1903, LAURIER'S GOVERNMENT UNDERTOOK the National Transcontinental Railway project. This was to be a government-built railway from Winnipeg through Northern Ontario to Moncton, via Quebec City. The Grand Trunk Railway had also committed to complete its line from Winnipeg to the west coast. The project's goal was to provide Western Canada with a direct rail connection to Canadian Atlantic ports and to open up and develop the northern frontiers of Ontario and Quebec. It would compete with the Canadian Pacific and the Canadian Northern Railways. A bridge near Quebec City was now seen as a national necessity.[1]

The financial condition of the Quebec Bridge Company was such that it could not possibly complete the undertaking without the aid of the Dominion Government. In light of the Transcontinental initiative, the federal government could have easily taken over the property from the Quebec Bridge Company and completed the bridge as a public work. However, Parent and his board of directors went to work and somehow convinced the government to fund their project.

With assurances from Ottawa that funding would be in place before the end of the year, Parent wrote to the Phoenix chief engineer in May and asked him if he would be prepared to travel to Ottawa to appear before a government committee to answer questions regarding the superstructure for the bridge. Deans wired Parent that he would go to Ottawa the next day, or any other day for that matter. Later that year, Deans and the substructure contractor, Davis, travelled to Ottawa to appear before government officials to answer questions as to the reasonableness of the cost of construction for the substructure and the superstructure.

In anticipation of favorable legislation, articles of agreement for the main span were prepared and signed by the Company and Phoenix on June 19, 1903. The ten-page contract was approved by the board the same day. David Reeves, president of Phoenix, attached a letter to the executed agreement, however, stating that it would not come into force until the proposed legislation was actually passed by Parliament. He did agree to continue working on the strain sheets and drawings as soon as Phoenix received the revised specifications, as approved by the government engineers. Reeves' conditions were accepted by the Quebec Bridge Company. The contract between Phoenix and the Company named Hoare as the chief engineer and Cooper as consulting engineer, but there was nothing in the document that clarified whose decisions had priority.[2]

Almost four years after his initial involvement, Cooper received word from Parent that the financial affairs of the Company were in order, and the work could now proceed. Cooper then "took up again, with the Phoenix Bridge Company and with the chief engineer, the necessary modification of the loads and stresses to suit a bridge of this magnitude." Although Cooper discussed his proposed changes to the specifications with Phoenix's chief design engineer, Peter Szlapka, he did so only to get the benefit of his views, not to get his approval or that of Phoenix. Deans and Szlapka encouraged Cooper to set aside the government's specifications altogether and substitute them with Cooper's own specifications, titled "Cooper's Specifications for the Superstructure of Railroad and Highway Bridges." But Cooper ignored the request; he would work with the government's specifications and modify them as he saw fit.

In June, Cooper sent the Quebec Bridge Company his proposed amendments to the government's 1898 specifications, which had been drafted by Hoare, based upon the Douglas 1896 specifications. The Company's original specification was, in Cooper's view, "a scissored one, not drawn upon any theory by any person having the importance of this bridge structure in his mind." He pointed out that, "Although a specification for a Canadian bridge, there was no recognition for the snow weight that must at times come upon this structure. The requirements for the wind strain are those practically imposed upon the Forth Bridge against the protest of the chief engineers of that bridge, Messrs. Baker and Fowler. The train load and train requirements are not as great as they should be. A large amount of the material in this bridge is going to be devoted to giving it horizontal strength against an imaginary and impossible wind, material that could be much more favourably placed to give the bridge vertical strength under higher train loading."[3]

Cooper proposed to "Correct the specifications to provide for less wind strain than that originally required, with a greater vertical loading than that at first required." Aware of the need to remain within the budget, Cooper wrote, "Being impressed with the necessity of restraining the weight of the structure under these new loadings and changes of loads so that it will not exceed the original estimated weight contained in the contract, I made modifications in the unit strains to be employed upon the various members, with a view of keeping the final weight within the limitations and yet obtain more harmony in the relative strength of the different parts of the structure." He assured the Company that they would "get a better bridge without increasing the estimated weight" and cost.[4]

In his engineering analysis, Cooper referred to the only comparable cantilever bridge in the world, the Forth Bridge in the east of Scotland, with a clear span of 1,710 feet. The English designers, Baker and Fowler, had used a dead load of 22,400 pounds per square inch, and two-thirds of that number for the live load. That meant that the working strain was about 20,000 pounds per square inch compared to the 21,000 pounds aimed at for the Quebec Bridge. After considering a number of factors, including the camber requirements, Cooper felt that the strains he had adopted for the Quebec Bridge were within the strains that were employed for the Forth Bridge. He amended the specification and increased the strain to 24,000 pounds per square inch, which was three-quarters of the elastic limit of 32,000 pounds. Cooper anticipated that the actual strain would never exceed 21,000 pounds. While acknowledging that such unit stresses exceeded the then-accepted practice in bridge design, Cooper defended his decision: "This is an exceptional bridge of exceptional length, and high strains are justified because the greater weight is that due to the weight of the structure itself, and any small uncertainty in regard to the live load would be comparatively a minor factor." These were to be the highest stresses ever used in bridge design and construction.[5]

Cooper's views regarding the design of the bridge in Scotland were made very clear at a meeting he attended in 1891 of the Engineering Society of West Pennsylvania. His criticism was unsparing; Cooper told his audience, "You all know about the Firth of Forth Bridge—the clumsiest structure ever designed by man, the most awkward piece of engineering in my opinion that was ever constructed from the American point of view. An American would have taken that bridge with the amount of money that was appropriated and would have turned back fifty percent to the owners, instead of collecting, when the bridge was done, nearly forty percent in excess of the estimate."[6]

Prior to submitting his proposed amendments to the board, Cooper met with Phoenix's chief design engineer, Peter Szlapka, to talk over the specifications for the bridge. Szlapka recalled later, "Mr. Cooper impressed upon me the importance of strictly following the revised 1898 specifications, but at the same time, to be prepared to consider special important features with him, irrespective of the requirements of these specifications. In view of Mr. Cooper's proposition to use, for certain combination of conditions, unit stresses as high as 24,000 pounds per square inch, or three-quarters of an average elastic limit of 32,000 pounds, I mentioned to Mr. Cooper the fact that a German professor proposed to use a fraction of the elastic limit for unit stresses for truss members, after first allowing for irregularity of shop work, for imperfect erection, for flaws in materials, etc." Cooper rejected Szlapka's suggestion; the structure was to be designed utilizing a unit stress of 24,000 pounds per square inch, without allowing for any of the unknown factors suggested by Szlapka.[7]

The board of the Quebec Bridge Company accepted Cooper's recommendation to amend the specifications without question, and they sought the government's approval, as required by the subsidy agreement. The proposed amendments were submitted to the deputy minister, Schreiber. Without looking at them, Schreiber turned them over to Douglas, the department's chief bridge engineer. It had been Schreiber's practice to rely on Douglas. "Nothing was approved by me that Mr. Douglas, after going through the figures, would recommend should not be approved."

Except for his initial involvement with Hoare in putting together the specifications for the call for tenders, Douglas had had no involvement with the tenders for the superstructure. It was only through hearsay that he learned that Cooper had endorsed the Phoenix tender and recommended their plan. In the spring of 1901, Schreiber had instructed Douglas to visit the site and examine the work that was being done on the piers for the substructure. Douglas went to the site periodically to inspect the piers and provide estimates of the work done for payment purposes. At one point, Schreiber had called on Douglas to travel to the site to deal with

SPECIFICATIONS AND FINANCING

an issue that had come up regarding the foundation for the south main pier. He travelled to Quebec City to meet with Schreiber and Cooper, who had travelled from New York. By the time Douglas arrived, the foundation issue had been resolved by Cooper and Schreiber, without input from him.

Douglas was, however, directly involved with the design and construction of the two 220-foot approach spans on each side of the river. Phoenix had submitted their design for these simple span structures directly to Douglas's department for approval, since they'd been designed and built before Cooper's amendments to the specifications came into effect. Douglas made changes to the floor design and the spans were built accordingly. When Cooper became aware of the changes, he complained to Hoare that they were eighteen to twenty percent heavier than the best requirements of the Pennsylvania Railroad and all first-class railroads in the U.S. He pointed out that for every pound put into the floor system, four to five pounds of extra metal would be required in the trusses to carry it and that this excessive requirement would "render it impossible to build the structure within the limitation of the financial ability of the Company." The design for the floor of the main span of the bridge, based on Cooper's specifications, would be much lighter.[8]

On June 29, 1903, the president of the Quebec Bridge Company wrote to the Minister of Railways and Canals in Ottawa with an unusual request that had come from the Phoenix chief engineer, Deans. Parent pointed out that the design and drafting work for the superstructure would take months to complete before any fabrication could begin. He wrote, "It is absolutely imperative that the continuous flow of the working drawings to the shops shall not at any time be interrupted, as the slightest delay will most assuredly lose a season's erection." To that end, he asked that the normal requirement of having the plans approved by the Department of Railways and Canals be waived, and that all specifications and designs signed by Cooper be accepted by the government.[9]

Schreiber, as deputy minister and chief engineer for the Department of Railways and Canals, reacted strongly to this request. He advised Parent that he was bound by the Department's standard specifications to review and approve all plans and specifications. He was aware, however, of the need for expediency and undertook to set up a process that would minimize any delays. Deans was not happy when Parent told him of Schreiber's response, and he wrote to Cooper expressing his dismay; "It certainly has proven to be a thankless task so far in trying to save the

Quebec Bridge Company a large amount of money without in the least affecting the efficiency of the structure."[10]

During this time, Douglas had concluded his review of Cooper's proposed amendments to the specifications and submitted his report to Schreiber. He recommended the adoption of some of the amendments, but he criticized the high unit stresses being proposed by Cooper and the suggestion to use the bridge for heavier rolling loads. He also recommended that the Quebec Bridge Company be required to submit new specifications and not merely amendments to the 1898 specifications. In Douglas's view, the unit stresses should be no greater than sixty percent of the elastic limit of medium steel, as specified in the department's 1901 general specifications for steel bridges. This was a new specification that would have applied to this bridge. The 1901 standard set the elastic limit for steel at 33,000 pounds per square inch; sixty percent of this limit (19,800 pounds) was to be used as the limiting stress for tension members and fifty-five percent (18,150 pounds) for compression members.[11]

Douglas was also aware that the American government had, in similar cases dealing with large structures, appointed four or five engineers to consider and determine unit stresses of unprecedented magnitude. He thought that "This matter was too important to be left to the judgment of Mr. Cooper," and he discussed adopting this approach with Schreiber. Anticipating that Cooper's amendments would be rejected, he even wrote to several prominent American engineers.[12]

On receiving Douglas's report, Schreiber had a tough decision to make; he was faced with having to choose between Cooper and Douglas for technical advice on this bridge. He decided on Cooper and accepted his amendments to the specifications. He did, however, accept Douglas's recommendation to engage other engineers to conduct design reviews. Schreiber proposed to hire a specially chosen bridge expert who would be an employee of the department and who would report directly to him. In his report to the Minister of Railways and Canals, Schreiber expressed his high regard for Cooper's professional standing; "That gentleman being a man of repute and reliability, his modifications may, therefore, reasonably be considered to be in the best interests of the work." But Schreiber closed with the following recommendation: "The department be authorized to employ a competent bridge engineer to examine from time to time the detailed drawings of each part of the bridge as prepared, and to approve of or correct them as to him may seem necessary, submitting them for final acceptance to the chief engineer of the Department of Railways and Canals."[13]

Shortly after submitting his report, Schreiber met with the minister of Railways and Canals and received approval for the changes. The minister reported to the government, and on July 21st, an Order in Council was passed, accepting Schreiber's recommendation. Schreiber was also instructed to hire a competent bridge engineer, as he suggested. His department communicated with a bridge engineer from New York who also had standing in the profession. They had progressed to the stage where they were discussing terms. Schreiber wrote to Cooper at this time, enclosing the order in council.

When Cooper read it, he was outraged, and he immediately wrote to Hoare. "I am in receipt of papers from Mr. Schreiber which surprise me. He is to select an engineer in New York who will examine from time to time the plans, approve or correct the same as to him may seem necessary. This puts me in the position of a subordinate, which I cannot accept."[14]

Cooper also wrote to Schreiber and Deans. When Deans read Cooper's letter, he too was incensed; not only was the Department of Railways and Canals going to review the plans, but a new additional review would take place prior to that. His plan had backfired. He replied to Cooper, "It would have been much better to have had Douglas as originally proposed rather than to have the present plan." Deans wrote to Hoare, "This would bring matters to a standstill, pending the appointment of such an engineer. As it now stands nothing can be done on plans." The work in the Phoenixville design office came to a halt. Deans also sent a letter to Schreiber urging him to abandon his proposed plan, and reminded him that his sole purpose for suggesting the Order in Council, through Parent, was to give Cooper the final authority to settle all details, the government approval being a mere formality, and in this way, save valuable time.[15]

Several days later, after Cooper had calmed down, he sent a note to Hoare, telling him that it was not his intention to circumvent or avoid the supervision of the government authorities. He pointed out, however, that the detail specifications that apply to ordinary bridges must be modified to adapt them to a work of this magnitude. "As the larger part of the rules in existing specifications are taken from my specifications, I know their inapplicability, unmodified to a structure of this magnitude. As the various members of this bridge will exceed anything heretofore made and will tax to the utmost the manufacturing appliances of the time, there should be given to the consulting engineer latitude to decide each case as it comes with promptness. The work would be delayed beyond reason if each case must be discussed, and consent given beforehand." He added, "My *chief* interest in this work is to obtain a work, which I can feel will crown my professional career of over forty years." Cooper underlined the word "chief."[16]

By August 13th, the situation hadn't changed. Cooper travelled to Ottawa to meet with Schreiber. He argued that the interests of the Company and the Dominion of Canada were identical in every respect and that he, Cooper, should have the final say. Schreiber succumbed to Cooper's will. He wrote to the minister stating that Cooper was "a man whose ability was never questioned, and whose experience in connection with bridge construction has been large; it was thought better to rely upon him rather than interfere with what he might do, what advice he might give."[17]

The bill amending Schreiber's recommendation was passed in Parliament on August 15th without debate, during the last hours of a session that had lasted eight months. The Order in Council directed that "The minister further represents that the chief engineer has this day reported, stating that, as the result of the personal interview had with the company's consultant engineer, he would advise that, provided the efficiency of the structure be fully maintained up to that defined in the original specifications attached to the Company's contract, the new loadings proposed by the Company's consulting engineer be accepted and that all plans be submitted to the chief engineer of the Department and until his approval has been given, not to be adopted for the work. The minister recommends that authority be given for following the course so advised by the chief engineer, the order in council of the 21st July last to be modified accordingly." Cooper's amendments to the specifications and his letter of June 2, 1903, were attached to the order.[18]

The specifications for the longest clear span bridge in the world were now left entirely in the hands of Cooper, subject only to the approval of the government authorities, which was now nothing more than a rubber stamp. Cooper now had an absolutely free hand, just as he'd intended. He wrote to Hoare: "I think under fair and broad-minded interpretation, this will allow us to go on and get the best bridge we can, without putting metal where it will do more harm than good."[19]

From that point onward, all plans bore a note: "According to the specifications of the Quebec Bridge Company as amended by Theodore Cooper." When plans arrived at Douglas's office, the first thing he and his staff would do was to check for Cooper's signature; that was "the principal part of it," according to Douglas. They would simply confirm that the plans were in accordance with Cooper's specifications attached to the contract. Although the stress sheets would arrive after the plans were submitted, it made no difference since no calculations were performed by Douglas's staff. The plans would be signed as examined and sent to the chief engineer. The plans from which the structure was built were all signed by the deputy minister and chief engineer of Railways and Canals, Collingwood Schreiber. The first plans prepared in accordance with Cooper's specifications arrived at the government offices on October 3, 1903.

By act of Parliament, on October 24, 1903, the government undertook to guarantee the principal and interest of the Quebec Bridge Company's bonds, up to a fixed amount of $6,678,000, such bonds to be payable in fifty years, bearing interest at the rate of three percent per annum, with a first charge secured by mortgage on the Company's franchises, tolls, and property. The government bonds were deposited with the Bank of Montreal, and advances were to be made by the bank to the Bridge Company as the need arose. One of the conditions set out in the agreement was that before the guarantee should be given, the Company was required to procure the subscription and full payment in cash of $200,000 of additional stock. A certificate was issued by the Quebec Bridge Company on February 26, 1904, signed by the secretary, Barthe, and sealed by the president, Parent, confirming that the additional stock to the extent of $200,000 had been duly subscribed and paid in full. These subscriptions bore interest at five percent per annum and were subject to a bonus of ten percent.[20]

The Company was now in a financial position to enter into contracts for the construction of the bridge under its new name, the Quebec Bridge and Railway Company. In February 1904, Parent wrote to Reeves advising him that the Company's finances were now in "satisfactory condition" and that Phoenix could now feel confident in proceeding with the work, knowing that payment was certain. Reeves replied on March 15th and advised that they were proceeding with the work "vigorously." The final agreement provided that Phoenix was to "construct, deliver and erect in the most substantial and workmanlike manner, to the satisfaction and acceptance of the consulting engineer and the engineer of the Company, and in accordance with the general plans and specifications attached hereto, the metal superstructure." The superstructure was to be completed by December 31, 1908, failing which Phoenix would pay liquidated damages of $5,000 per month.

Quebec City was founded by Samuel de Champlain on behalf of King Henry IV of France on July 3, 1608. The Phoenix Bridge Company was told unofficially by Parent that he fervently hoped the bridge would be completed in time for the Tercentenary celebration. The Canadian government expressed a similar desire. If Phoenix was going to finish the job by the middle of 1908, they would have to accelerate the work. They considered the possibility but made no promises.

CHAPTER 6

DESIGN

THE PLANS FOR THE BRIDGE had gone through several stages of development since the original Phoenix proposal of November 1897. Changes were made to incorporate the longer span in 1901, but minimal work had been done by Phoenix since then, except for a few minor design details. Following receipt of Cooper's revised specifications in 1903, and the government's financial commitment, work began in earnest on what was to be the longest clear span bridge in the world.[1]

Peter Szlapka was the chief design engineer for Phoenix. A graduate of the Royal Polytechnic School in Hanover, Germany, Szlapka had started with Phoenix in 1880 as a bridge draftsman. Six years later he'd moved to the design department and had been there ever since. Szlapka had never worked in the field; he was strictly a designer. Prior to the Quebec Bridge, he'd designed six large bridges for Phoenix, but nothing compared to this—the longest span he'd ever been involved with was a cantilever bridge over the St. Lawrence River at Cornwall, Ontario, with a span of 840 feet; but that bridge weighed only 1,200 tons. The clear span for the Quebec Bridge was more than double that and the steel superstructure was over twenty-five times heavier.[2]

As far as Cooper was concerned, when it came to Szlapka, he had "implicit confidence in the honesty and ability of Mr. Szlapka, and when I was unable to give matters the careful study that it was my duty to give them, I accepted the work to some extent upon my faith in Mr. Szlapka's ability and probity." In turn, Szlapka viewed Cooper as "the highest type of an able and honest engineer. Considering Mr. Cooper one of the ablest and most experienced bridge engineers in the country, I discussed fully with him all main features of the bridge. His advice and directions were always sought and appreciated." Szlapka visited Cooper in New York at least once a month. Cooper visited Phoenixville three times during the entire period of design and fabrication of the bridge. He never visited the bridge site during the erection of the steel.[3]

Szlapka had two subordinates who in turn had four or five assistants each, and they were supported by two dozen draftsmen. Charles Scheidl was Szlapka's chief of detailed design and had been since 1889. After Phoenix received Cooper's revised specifications in 1903, the first thing Scheidl did was to find a quiet office to go over the outlines of the bridge and the general stress sheets. Determining the length of all bridge members was going to be a challenge as none of the trusses had a single horizontal member; they were all inclined at various angles, owing to the curve of the structure.

Scheidl realized that the dimensions of the compression members were such that transportation could prove to be a limiting factor. The carrying loads and clearances of the different railway companies transporting material to Quebec had to be studied before any of the large compression members could be designed. Sketches giving the extreme dimensions and weights had to be prepared for the different railroad companies. These companies had not only to determine if these large pieces could be shipped over their own lines, but also whether they could be shipped the whole distance, from shop to bridge site. The rail cars also had to be specially rigged to carry these pieces; measures had to be taken to ensure the proper distribution of the loads between sets of wheels on the cars.

The cantilever principle originated in the Far East. It comprises a rigid structural element, such as a beam, anchored at one end to a support from which it protrudes. A cantilever allows a structure to overhang without a support at the other end, similar to a diving board. Construction of a cantilever bridge is carried out from both shores, either simultaneously or in sequence, and the two sides meet in the middle. This type of bridge requires no temporary shoring in the river channel during construction, a critical factor for the Quebec Bridge, since navigation in the St. Lawrence River could not be disrupted. For this bridge, the south arm was to be built first and be substantially completed before work would begin on the north shore so that the cranes, tools, and scaffolding could be re-used. Geographically, the north shore is on the Quebec City side of the river and Lévis is on the south shore. Quebec City is seven miles downstream, to the east of the bridge, and Montreal is 160 miles upstream, to the west of the bridge. As you look from the south shore of the river to the north shore, Montreal is on your left and Quebec City is on your right.[4]

There were two piers on each side of the river, an anchor pier and a main pier. The main pier was the pier closest to the water and carried most of the weight of the bridge. The anchor pier, which was farther away from the river nearer the river bank, was just that—it acted as an anchor for the cantilever, which prevented the bridge from tipping over into the river, similar to the end of a diving board that is anchored to the floor. The steel superstructure above the piers consisted of a truss on each side of the bridge deck. These two trusses were tied together with lighter structural bracing, running overhead and under the floor.[5]

The total length of the bridge was 3,240 feet, comprising two approach deck spans of 220 feet each as well as a 2,800-foot cantilever structure, which was made up of two 500-foot anchor arms, two cantilever arms of 562-1/2 feet each, and a 675-foot long suspended span in the middle. The clearance above high-water had to be 150 feet for a minimum distance of 1,200 feet. The main towers or posts, resting on the main piers, were 315 feet high and rose 400 feet above high-water. As you looked down the center-line of the bridge, the trusses on each side of the bridge were spaced sixty-seven feet apart, wide enough to accommodate two sets of railway tracks, two sets of streetcar tracks, and two roadways. Sidewalks on each side of the bridge were to be carried on short cantilever beams extending outside of the trusses. The enormous dimensions made the bridge a massive structure in every sense.[6]

The anchor arm is the section between the anchor pier and the main pier on the shore of the river. The anchor arm transferred the upward forces generated by the weight of the cantilever arm and suspended spans that extended out over the river, onto the anchor pier which held it all down. The cantilever arm was attached to and supported by the anchor arm, and one half of the suspended span was attached to and supported by the cantilever arm. Once the two structures came together in the middle, they would provide a clear span of 1,800 feet between the centers of the main piers, thereby exceeding the span of the Forth Bridge by ninety feet, a new world record for any type of bridge.

The anchor arm was subdivided by vertical members into ten frames, or panels, of fifty feet each. The cantilever arm also consisted of ten panels, each one measuring fifty-six feet, three inches, as did each of the six panels comprising one half of the suspended span. The spans from each shore were identical. This meant that the members for each side of the bridge could be fabricated at the same time, to ensure as close a match as possible.

The Forth Bridge in Scotland had been built piece by piece, almost entirely on-site. Each member of the steel superstructure was assembled in place; this was the English and European method of building such steel structures. American bridge-building, on the other hand, differed in that as much work as possible was done in the shops, and the assembled pieces were shipped to the site where the field crews erected and joined them together, first with bolts and finally with rivets. The primary reason for the different approaches was the cost of labour; wages were much higher in the U.S. and Canada compared to overseas.

In general, the trusses on each side of the bridge were pin-connected, employing rows of eyebars for its top chord chain and all other main tension members. The compression members on the bottom of the trusses generally had riveted connections, but the posts were pin-connected to the bottom chords and the latter were pin-connected to the shoes on the piers, resting on a 24-inch diameter pin that carried the main post. The use of pin connections was simpler and cheaper, but less safe than riveted connections; this was consistent with the American approach to bridge-building.

The top chords of the trusses on each side of a cantilever bridge are in tension and the bottom chords are in compression, as are the main vertical posts. The Phoenix design differed from most cantilever designs in that the top and bottom chords for the anchor and cantilever arms were curved. It was common practice for these chords to be straight, making it easier to fabricate and erect. This curvature was there solely to give the bridge a more pleasing appearance. The top chords for the anchor and cantilever arms of the Quebec Bridge were made up of dozens of parallel steel eyebars. Measuring fifteen inches deep, two inches thick, and reaching lengths of more than eighty feet, they were forged with flared ends and drilled for their pin connections.

The bottom chord members, which were the most heavily loaded compression members in the structure, were massive rectangular sections sixty-eight inches wide, fifty-four inches deep, and measuring fifty to fifty-seven feet long, depending on where they were in the arch of the curve. These built-up members weighed as much as 100 tons each. The chords themselves were made up of four vertical webs, or ribs, with each rib consisting of four one-inch steel plates stitch-riveted together to from a single built-up plate, four inches thick. Each of these ribs was finished with an angle, or flange, riveted to the top and bottom of the rib.[7]

The means adopted by the designers to stiffen and bind together these four massive ribs, so as to act as a single member, was by shop-riveting lattice bars across and diagonally on the top and bottom flanges of the ribs between the end splice plates. The lattice bars were made of 4 x 3-inch angles, weighing eight and a half pounds per foot. The theory of columns was not developed at that time to provide empirical formulas for designing such lattice bars; the designer was left to draw upon his own experience since no relevant testing had been performed. Cooper's own general standards for bridges were not overly helpful either; they stated simply, "The size and spacing of the lattice bars should be duly proportional to the size of the members." Compared to the size of the ribs themselves, these angles appeared light for the task assigned to them. Cooper had, on prior occasions, designed large columns such as these with a heavy cover-plate on one side, thoroughly riveted to the various web flanges, with latticing being used on the other side of the member only. Szlapka's design for the Quebec Bridge did not call for such a continuous cover-plate; latticing was used on both sides.

The flanges on the two outside ribs were wider and therefore the lattice angles could be attached with the use of two rivets. The inside ribs had smaller flanges that could only accommodate a single rivet. This critical feature of the design would prove fateful.

The four deep and narrow ribs were finished with square ends so that the load could be transmitted from one chord to the next by means of contact of the abutting ends. The chords were to be joined one to the other at their ends by means of eight splice plates. There was an upper and a lower horizontal plate, and six vertical plates, one on each side of the two outside ribs, and a single vertical plate on each of the two inside ribs. The order of erection required that the lower horizontal plate should be put in position before the next chord was set; the vertical plates were added next, and the erection of the joint was finished by bolting on the upper horizontal plate.[8]

The steel for the superstructure would comprise ordinary structural steel, with a low nickel content—there is a direct correlation between the percentage of nickel or high carbon steel used to the strength of the steel. The steel for the Manhattan suspension bridge that was being built at about the same time by the Phoenix Bridge Company utilized high carbon steel with fifty percent nickel content.[9]

After the first sections of the lower compression chords of the anchor arms were fabricated in the shops, the president of the Phoenix Bridge Company, David Reeves, told Szlapka that two of the shop foremen had mentioned to him that they thought the lattices on the chords appeared too light, and that they were liable to be injured or damaged in handling in the shops or during transportation to site. Szlapka replied that lattices of any size might be injured and destroyed if carelessly handled in the shop or during transportation. When Szlapka later reported his conversation to Cooper, the consulting engineer told Szlapka that he had looked into the question of the strength of the lattices while checking the plans and that "we had it all right." Cooper also remarked that he had no doubt that "Mr. Reeves would be very glad to increase the tonnage," since Phoenix was being paid by the pound.[10]

Phoenix submitted their plans to Cooper for the top tension chord eyebars of the anchor arm in June 1904. Cooper had previously met with their designers and told them he would not accept bars exceeding two inches in thickness; his experience had proved to him that when that thickness was exceeded, satisfactory bars couldn't be fabricated. In their plan for the top chord, the Phoenix engineers had used bars two and a half inches thick and arranged them at angles, which was thoroughly unsatisfactory to Cooper. He called for a new design. Szlapka travelled to New York and told Cooper that he'd had his best men on it for two months and this was the best they could do. Cooper told him he would never approve it and finally he was "compelled personally, although it was work I had not done for twenty years, to redesign the whole system. It was very arduous and trying work and when I was through, I was thoroughly exhausted." Phoenix worked with Cooper's design and the thickness of the eyebars never exceeded two inches. Seventy-three full size tests were made on the eyebars between 1904 and 1907, as required by Cooper's specifications. The eyebars would prove to be the strongest part of the entire structure.[11]

Cooper's own standard bridge specifications required that tests also be conducted on the compression members. The Phoenix Iron Company owned the most powerful machine for compression testing in the world, but Cooper never ordered such tests as the Phoenix machine simply wasn't large enough to accommodate a

DESIGN

full-size compression member. It would have been possible to fabricate and test a scaled-down version of a compression member, but this was never proposed by anyone or requested by Cooper. No tests of any kind were conducted on what was considered to be the weakest part of the structure.

Bridge design is an iterative process. For all bridges, nearly all of the dead load to be carried is the weight of the structure itself. The engineer designing the bridge has to begin by making assumptions as to what the weight of the bridge itself will be. He then adds those loads prescribed by the specification for traffic, wind, and snow, using his experience to estimate the size and weight of the individual parts of the bridge that will be required to carry these loads. From that number, he can calculate the stresses on each member. This in turn determines the cross-sectional area of steel required to carry the load, which dictates the size of the member. After the initial design is complete, the engineer adds up the total weight of the steel members and their connections and repeats the process to see whether the original assumptions were correct. In bridges of ordinary design and dimensions, experience enables the designer to estimate the weight so accurately that a recalculation of the weight is not always necessary. The unique character of the Quebec Bridge, however, with its length of span and high unit stresses, demanded that this iterative process take place probably several times. The contract between Phoenix and the Company also required such a recalculation. Cooper's own specifications stated that "before any construction of any part of the structure shall be proceeded with, complete working drawings shall be furnished, showing all details of construction."

Phoenix had re-estimated the weight of the superstructure after the length of the span was increased to 1,800 feet, and they revised their estimate again after Cooper reviewed the stress sheets for the suspended span. In April 1904, Szlapka advised Cooper that the weight of the bridge would not be more than five percent above 31,000 tons.

The Phoenix engineers had planned to complete the shop drawings for the anchor and cantilever arms by August 1904 to give the shops eight months to fabricate the materials, thus allowing erection to begin in May 1905. In accordance with company policy, Deans and Superintendent Reeves instructed Szlapka to "place with the shops any shop plans as soon as approved, and to generally arrange the office work so as to ensure continuous working on the bridge, in the shops and in the field." Szlapka had his engineers concentrate their efforts on the anchor arm since it was to be erected first. The design details for the anchor arm and the majority of the cantilever arm were completed in 1904. Cooper approved the details for the anchor arm in October. The shops in Phoenixville commenced steel fabrication soon after, before the overall design of the entire bridge had been completed.[12]

Szlapka had confidence in his assumed weights. In his view, and that of his superiors, it wasn't practically possible to prepare an initial design of the entire bridge, recalculate the weights, and go through another iteration. Phoenix believed that its broad experience enabled it to proceed in this manner, and besides, the tight schedule for the Quebec Bridge precluded them from doing so. As a result, the Phoenix engineers did not recalculate the actual weight of the structure.

When the drawings for the anchor arm were almost complete, only a very small portion of the work had passed through the shops, and erection at the site had not yet begun. Nevertheless, Cooper approved the detailed design for the anchor arm. He did so fully aware that the detailed design of the cantilever arm and suspended span was not yet completed, which meant that the weight of these parts of the structure, which the anchor arm would carry, was still unknown. In order to get things underway, the engineers, including Cooper, one of the foremost bridge engineers in America, ignored their duty to re-calculate the weights, and blindly accepted Szlapka's assumed weights for the structure. Szlapka had simply followed the company's business practices, which made no allowances for bridges of unprecedented dimensions, such as the Quebec Bridge.[13]

Eighteen months later, in February 1906, Cooper and the Phoenix engineers learned that the weight of the bridge members that had been fabricated and shipped to the site far exceeded the weights that had been assumed for the design. The chief shop inspector for the Quebec Bridge Company at Phoenixville reported that the actual weights of the cantilever arm and suspended span were twenty percent higher, and that the anchor arm was thirty-six percent heavier than what had been assumed for the design. The error, or omission, wasn't that the assumed weights were incorrect (that is more or less unavoidable), but that a re-computation hadn't been carried out to more accurately reflect the actual weight. This same mistake was made for the design of the cantilever arm and for the suspended span.[14]

By this time, all of the south anchor arm, the center posts, and two panels of the cantilever arm had been fabricated, and six of the ten panels of the anchor arm had been erected at site. Szlapka travelled to New York to tell Cooper. Cooper understood immediately. He sat down and estimated the increased strain on the structure resulting from the higher than assumed weight; it was roughly seven per cent. The two engineers recognized the gravity of the situation. They had a choice to make: either ignore it or re-design the bridge using the actual weight, in which case all of the work at site would have to be dismantled, all the steel already manufactured would have to be scrapped, the project would suffer significant delays with the resulting financial consequences, and everyone's position would be at risk. Cooper looked at Szlapka and said, "We will have to submit to it." In Cooper's view, it was still safe to proceed. He said what everybody wanted to hear, and so they all submitted to it, including Hoare, the chief engineer of the Quebec Bridge Company. The project continued, uninterrupted, without any changes being made to the design.[15]

Cooper and the Phoenix design staff then re-calculated the weight for the remainder of the structure and found that the total weight of the steel would be more than 36,000 tons, almost twenty percent more than what was originally assumed. When final weights were calculated a year later, this figure had grown to almost 39,000 tons—the bridge was thirty percent heavier than what it was designed for. Cooper's increased stress of seven percent was now ten. This weight error, compounded by the unprecedented allowable stresses that had originally been adopted for the design, was sufficient to have condemned the bridge. If the bridge had been completed as designed, the actual stresses would have been much greater than those permitted by the specifications. In spite of this, no changes were made, and the work continued—the schedule demanded it.[16]

Once design drawings were completed, they were forwarded in duplicate to the consulting engineer for his approval. Once approved, seven additional prints of complete, checked drawings were sent to him for his signature and six copies were returned by him to the Phoenix Bridge Company. Five copies were then sent to the chief engineer of the Quebec Bridge Company, Hoare, who forwarded these to the Department of Railways and Canals. Phoenix received one print back, approved by the Dominion Government. Occasionally, Phoenix didn't wait for Cooper's approvals before proceeding with fabrication in the shops, "in order to ensure continuation of the work in the shops and in the field." They often proceeded without the necessary approvals from the Department of Railways and Canals, which they viewed as a mere formality.

Hoare's inspector in Phoenixville told him that Phoenix had Cooper's consent to roll a limited quantity of steel for the sections that he had approved, entirely at the risk of Phoenix. If they weren't approved, and changes were necessary, then Phoenix would bear the cost. Of course, nothing was ever rejected; Cooper's approval was, in reality, the only one required. Hoare made sure, however, that his inspector omitted from the monthly reports the metal rolled ahead of the certified plans, thereby delaying payment for these sections until the Department's approval was in place.

Cooper and his meager staff of one checked the strain sheets and all detailed plans "at my own expense and it was not sufficient for the purpose, considering the other duties which were imposed upon me improperly." His annual fee had originally been set at $7,500, but he later cut it to $4,000 after seeing for himself the financial difficulties of the Quebec Bridge Company. This amount had to cover all of his expenses, including the wages of his assistant, Brent Berger. Cooper raised the issue with Hoare, but nothing was done. Cooper's desire to see the project through kept him from taking any action to address the fee issue.

In 1904, Parent was in Cooper's office in New York and he asked Cooper when he was going to Quebec next. Cooper replied, "Mr. Parent, I never expect to be able to go to Quebec again. I am under the ban of my physician, and I feel that I ought to be relieved of the responsibility which is upon me, as it is impossible for me to give it that attention that I conscientiously feel I should do."

Parent protested, "Mr. Cooper, we never intend to let you go until the bridge is done. We have confidence in you and we want your services continued."

Cooper also told Deans that he thought he should withdraw. "While I appreciated the complication that it would involve and the difficulty of their mutually selecting somebody who would be satisfactory, I would gladly withdraw from further responsibility."

Deans likewise protested and said they couldn't submit to that, since "they did not know of anyone upon whom they would be willing to submit an important contract like the one under execution."

Cooper relented. "Realizing this difficulty and feeling also a pride and a desire to see this great work carried through successfully, I took no further action."[17]

During the three years that it took to complete the detailed design, Cooper came up with numerous suggestions for change, many of which were accepted and implemented by the Phoenix engineers. At one point, Szlapka was meeting with Cooper in his office in New York, discussing a relatively minor change being proposed by Cooper. Szlapka didn't agree with the change, but not wishing to criticize Cooper's scheme directly, he stated that it might be criticized by the profession. Cooper replied, "There is nobody competent to criticize us."[18]

CHAPTER 7

INSPECTION REGIME

THE ORGANIZATION THAT WAS ESTABLISHED for inspection of the work in the shops and erection at the site was not in accordance with Cooper's advice. In his discussions with Hoare, he tried to impress upon the chief engineer of the Quebec Bridge Company the need for a rigid inspection regime in the shop that focused on the technical features of the plans, which in his view required technical engineers, possibly with shop experience. These engineers could then inspect the erection and after completion of the project, they could take charge of the maintenance and general supervision of the structure on behalf of the owners. To this end, he thought that "It would be fair and proper that these inspectors should be Canadians, graduates of Canadian institutions, because the men having charge of this work would have to live there, and they should be men of the country." Hoare didn't agree, and so Cooper "did not succeed in obtaining at that time the men I hoped for." Hoare and the Phoenix managers had their own ideas about who they wanted to act as inspectors.[1]

Hoare first met Robert Kinloch when he came to site in 1902 as an inspector on the approach spans for a testing lab out of Pittsburgh. Kinloch left a good impression with the chief engineer and Hoare asked him if he'd like to return in the future to work on the big spans. Kinloch said he would be glad to come back if he wasn't otherwise tied up some place else. From the time of completion of the approach spans until the superstructure work resumed in 1905, Kinloch worked in Omaha. The two men corresponded, and Hoare told Kinloch he would keep a position open for him. When he finished in Omaha, Kinloch wrote to Hoare, letting him know that he was available; Hoare told him to report to site. Kinloch arrived on Dominion Day, July 1st, 1905. The crews were erecting the falsework for the anchor arm—they hadn't yet started to raise any of the iron for the big spans.[2]

Kinloch had worked in bridge construction since the age of sixteen. Although he wasn't a trained engineer, he had twenty years' experience in the erection of bridges, mostly with the American Bridge Company, as well as with a number of railroad companies throughout the U.S. He had worked as a labourer, a pile driver, and as a general bridge man, an individual who is able to work as an erector, riveter, derrick operator, or any other function needing to be filled in bridge construction. He worked his way up and became an assistant foreman, a position he held for ten years. Eventually, he became an inspector, working closely with engineering departments. Since 1887, Kinloch had worked exclusively on large bridges.

After his arrival in Quebec, Kinloch spent most of his time in the storage yard getting familiar with the structural bridge pieces as they arrived; he wanted to learn as much as he could about the structure. It was his job to see that the iron was in good shape before it went into the bridge. His inspection took place just prior to erection. He hopped on the rail cars as they arrived from the storage yard at the bridge site and looked the members over carefully. Kinloch didn't take any measurements but he eye-balled each piece to make sure it hadn't been damaged since the last detailed inspection in Phoenixville. He also looked for the mark of approval left by the shop inspectors. The Quebec Bridge Company's inspectors at the Iron Works shop in Phoenixville inspected the material as it came out of the shop. Once a member was passed, the inspectors would indicate their approval by painting a white "Q" on the material and, with the use of a hammer with a "QB" stenciled on the end of it, strike the metal a blow to leave an indented "QB" mark on the steel. This was the final formal inspection of the steel members (unless a member was altered, in which case, it would be re-inspected). There were no other formal inspections of the members prior to or after transport to the yard in Quebec, or prior to the material being erected into the structure. Occasionally, Kinloch found that a member wasn't marked. He wouldn't reject these pieces, but he would spend more time examining them. He never found any significant defects, except for an occasional bent angle, which he marked to indicate it had to be straightened or cut off and replaced. Overall, he was impressed with the quality of the material that came out of Phoenixville.³

Although he didn't have the authority to stop the work—not without taking it up with Hoare or the Phoenix foreman—it was Kinloch who gave the go-ahead for any riveting, when the pieces would be permanently joined together. The riveting crews would not rivet anything together without Kinloch's approval. Kinloch knew and understood the work being done and was respected by the men. He was on the job during the entire erection of the south anchor span, except for about four weeks when he'd gone home twice to see his father; first when he got sick, and then when he died.

Norman McLure had graduated with a degree in civil engineering from Princeton University in 1904. While attending school, McLure had spent his summers working as a surveyor and inspector on bridges in New York. Following his graduation, McLure applied to Phoenix for a position as an engineer. Knowing that Cooper was looking for a young graduate with some experience in bridge work, the Phoenix chief engineer, Deans, turned his application over to Cooper. He wrote to him that "McLure has had just the experience which you desire for a man to be your representative Quebec field inspector." Cooper interviewed the young man in New York, found him satisfactory, and appointed him assistant inspector in the shops in Phoenixville, telling him that if he proved himself, he would be inspector of erection. Cooper asked the Phoenix chief designer, Szlapka, for his help in educating McLure for the position and to give him his assessment of McLure's capabilities. Szlapka later told Cooper that he found McLure energetic, very active, very bright, and thoroughly capable of understanding the work Cooper had in mind. Based on Szlapka's recommendation, Cooper later promoted McLure and advised Hoare of his decision.⁴

Hoare responded that he had already appointed Kinloch as inspector for erection; Cooper objected, telling Hoare he believed that Kinloch wasn't qualified to perform the duties that Cooper expected from this individual. Cooper was determined; he drafted a letter of instruction for McLure and told him to hand it to Hoare upon his arrival at the bridge site. The letter referred to McLure as "Inspector in charge of erection, Quebec Bridge." Cooper's instructions made it clear, however, that McLure was to report to Hoare. He was also instructed to watch and monitor the work, carry out any instructions given to him by the "chief engineer," Hoare, and report to Cooper once a week. In case of emergency,

McLure was to report first to Cooper and then to Hoare. The letter invited Hoare to add anything he thought necessary. When McLure showed up at site, he handed Cooper's letter to Hoare. Hoare had nothing to add and begrudgingly accepted McLure's appointment. Fortunately, McLure and Kinloch got along and worked well together. Kinloch took the young engineer under his wing. McLure's role was more technical, as compared to Kinloch's; he responded to technical questions as they arose and communicated with Cooper to get instructions.[5]

McLure felt he was responsible to both Hoare and Cooper, but when it came to technical matters, he relied on Cooper alone. He became Cooper's eyes and ears on the job—his written weekly reports were critical as Cooper never travelled the six hundred miles to Quebec during the two years of steel erection. It wasn't until the work of erection had progressed to the point where they had placed nearly all of the lower chord members of the anchor arm on the falsework that Hoare finally called on McLure for assistance. They were wedging out the lower chord for camber, something Cooper was convinced neither Hoare nor Kinloch understood, and that was the reason for calling on the young engineer. McLure proved himself by solving the problem with the camber; after that, Hoare realized that the young McLure could be useful and an asset to the job. McLure made a point of walking around the completed work on a daily basis, and not just from track level; he climbed on the top and bottom chords, transverse struts, and everywhere. McLure inspected all of the structural members with the exception of the first nine bottom chords of the south anchor arm; these had already been in place by the time he arrived at site. He reported to Cooper, Hoare, and the Phoenix supervisor when anything was amiss.[6]

In early 1902, the Phoenix chief engineer, Deans, received a letter from a professor of civil engineering at the Massachusetts Institute of Technology, suggesting the name of Arthur H. Birks as a desirable man for his engineering department. The professor wrote, "I have an exceedingly good man who graduated in architectural engineering and has been taking a post-graduate course with me. Birks is a man of exceptional ability in this line and having taken all my work in structures is as well up in bridge work as building work. He has also had some experience, having worked one summer with a bridge company, and one year in an architect's office. He is an exceptional man." Based on this recommendation, Phoenix hired Birks and he went to work in their design office as a structural draftsman. After six months, the Phoenix managers, Deans in particular, realized that Birks had traits of character and ability that would suit him for erection work in the field. He was transferred to the erection department and for the next two years worked on several bridge projects in the eastern United States. In early 1904 he returned to the Phoenixville office to work on the erection plans for the Quebec Bridge. Birks became familiar with every detail of the erection scheme and the expected behavior and movements of the trusses during erection. This, together with his thorough technical training and reliability, led to his being appointed by Deans as the Phoenix resident engineer of erection for the Quebec Bridge. Birks left Phoenixville for Quebec in September 1904.[7]

Cooper wasn't happy with Deans' decision. In his view, the inspector for Phoenix should be "fully cognizant of the details of the structure, the action of the different members under the different strains and camber movements and who would have the technical knowledge to take action if, at any time, the theoretical expectations should not be obtained, to determine why such result was not obtained and be able to direct the necessary corrections." Cooper didn't think Birks was this man. He felt Birks didn't have the training and knowledge to "fit him as the engineering representative of the contractor on such an important structure." In spite of Cooper's disagreement, Birks became Deans's man at site, his eyes and ears on the project. Like McLure, Birks would do nothing without first checking with his boss.[8]

Birks was the technical adviser to the foreman for the erection and handling of the individual structural pieces. Although Birks had no authority over anyone or gave any orders, as the technical adviser, he played a critical role in the erection of the superstructure. He was under strict orders to refer all but minor issues to the engineers in Phoenixville for their input, and he reported to and received verbal instructions directly from Deans. Birks saw to it that the erection was carried out in accordance with the instructions from the Phoenix office, independent of the foreman. Erection work did not proceed until Birks was satisfied.

The two young engineers, McLure and Birks, also got along well. They were tasked with checking and measuring all materials that arrived at the site before the pieces went into the bridge. They did this prior to the members being taken off the cars when they arrived at the bridge from the storage yard. They confirmed the identity and dimensions of the members and made sure they were properly assembled according to the erection plans. They also inspected the metal at the yard for anything that might have been damaged during transportation.[9]

INSPECTION REGIME

Survey control is a critical element in any type of construction, particularly on a project like the Quebec Bridge. Frank Cudworth was appointed by Phoenix as the resident engineer in charge of instrument work. The Phoenix foreman was Cudworth's boss, but he also took instructions from the Phoenixville office. A civil engineer from Dartmouth College, he'd worked in the U.S. Navy yard at Portsmouth before joining Phoenix. Cudworth provided survey control, making sure that the bridge was properly aligned horizontally and vertically. The field supervisors, as well as Birks, followed his instructions when it came to the positioning of the elements of the superstructure. Cudworth worked closely with Birks—they helped each other out. This was Cudworth's first bridge job, and one of his duties was to take and develop photographs as the work progressed.

Everybody got along, trying to make a success of the bridge, including Birks, McLure, Kinloch, and Cudworth. Kinloch said that Birks was the best man on erection he had ever seen; he could get plans on a moment's notice.

When it came to the appointment of inspectors in the Iron Works shops, Cooper once again wanted young men right out of college, men he could train according to his own ideas. The manager of the Phoenix Iron shops protested strongly, desiring more experienced men. Although Cooper had made it clear from the beginning that he would not assume responsibility for inspection of the steel in the fabrication shops, he did expect the senior inspector representing the Quebec Bridge Company to report to him directly. On one of his rare trips to Phoenixville, just when fabrication was getting underway, he met with David Reeves, president of the Phoenix Bridge Company. The two men discussed inspection in the shops and Cooper's preference for hiring younger men. Reeves told Cooper that the present inspector, Edwards, had been in their shop for years and that they considered him a very competent man. He urged Cooper to consider appointing him as the senior Quebec Bridge Company inspector.

Edwards had seventeen years' experience as an inspector on bridges and building work, primarily for the purchasers of steel, not the fabricators, which suited him well for the position. Cooper met with Edwards in New York and discovered that he had done some inspection work for him on another project, which Cooper had found satisfactory. He appointed Edwards to be the Quebec Bridge Company's inspector in the shops in May 1904. Although Edwards reported to Cooper as well as to Hoare, he remained an employee of the Phoenix Iron Works Company.

Cooper visited the shops just three times after the work commenced. He relied on Edwards, who made verbal and written reports to Cooper and visited him once a month. McLure spent his time in the shop under Edwards' supervision, before taking up his duties as inspector of erection at the site. Edwards was supported by one assistant, an experienced man like himself.

There was hardly ever occasion for Edwards to reject any materials from the shops as these were rejected by the manufacturers themselves. The eyebars for the Quebec Bridge were produced by the Central Iron and Steel Company from Harrisburg, Pennsylvania, as the Phoenix Iron Works Company lacked the capacity and the specialized manufacturing equipment needed to produce them. Their shop foreman showed Edwards, on one of his inspections, a huge pile of eyebar material that they had rejected themselves. As a result, only a very small percentage of the material submitted for inspection was ever rejected.

Edwards organized special tests to be carried out by Phoenix on some of the materials from the shop, primarily the eyebars. Cooper incorporated some of the test results in an article written by him and read before the American Society of Civil Engineers in 1906 under the title "Some New Facts About Eyebars." Owing to his involvement with the longest span bridge in the world, Cooper's reputation continued to flourish; he was breaking new ground.

When checking the materials prior to shipment, the inspectors in the shops measured the pieces with a special metal tape that had been calibrated with the master tape used by the shop. For long pieces, they used a scale and pulled on the tape with a force of ten pounds; one man would hold the scale and the other would read the measurement; they would then switch places and the other would take a measurement. Once they each had a measurement, they would then tell each other what their measurement was. Once they agreed, that number was compared to the specified dimension on the drawing. If they matched, it was recorded; if not, the pieces would either be rejected or altered, if possible.

Other than some minor painting issues, there were few complaints about the materials from the people at the bridge site in Quebec. There were also never any problems with the drawings. The Phoenix shop appliances used to handle the material were the best the inspectors had ever seen. The workers in the shop took great care to ensure nothing got damaged. Everyone had received special instructions to make this a master job. Inspectors from other projects had come

INSPECTION REGIME

to the Phoenix shop and commented that since the Quebec job had started, the materials they were getting for other work was better than ever.[10]

The Department of Railways and Canals had their own inspector in the shops in Phoenixville, but he was there solely to keep track of the individual pieces of the structure for payment purposes. Similar inspections were carried out at the bridge site by another engineer from the department, who visited the bridge at least once a month. For quality inspection, the Department initially relied on its chief bridge engineer, Douglas; he visited the site several times during the construction of the substructure piers. For the superstructure, however, Schreiber and his department were relying on Cooper, as he was, according to Schreiber, "really the man who looked after that. He was supposed to visit it frequently." But of course, Cooper never visited the bridge once during the erection of the superstructure, relying solely on weekly reports from McLure for his information. Schreiber retired in 1905, before the erection of the superstructure had started. Douglas never visited the bridge once after the piers were completed; he never saw any of the superstructure except for the approach spans. As far as the department's chief bridge engineer was concerned, this was Cooper's bridge and he could deal with it.

Nothing changed after Schreiber's retirement; his successor signed the drawings approved by Cooper but made no effort to ensure that the department was represented by a full-time resident inspector on the longest span bridge ever built.

Hoare was paid $150 per month for the first three years of his involvement with the bridge. His salary went up to $400 per month in 1900 and to $6,000 per year in 1905. Cooper received $4,000 per year for all of his services, which was to include the cost of his assistant, Berger. The Company appeared not to consider the discrepancy, or ignored it. Cooper said he raised it with Hoare, but nothing was done.[11]

CHAPTER 8

THE BUILDERS

THE ERECTION SEQUENCE FOR THE structure would follow the normal procedure for cantilever bridges; construction of the north and south arms would proceed outward from each shore towards the center of the river channel, but not at the same time. The southern arm on the Lévis side would be erected first, followed by the north arm on the Quebec side, thus allowing the builder to re-use the materials and equipment needed for erection on the opposite shore. The anchor arms on each shore, which spanned 500 feet between the main pier and the anchor pier, would be erected first with the use of temporary steel falsework or shoring. The falsework was designed so that the panel points where the bottom compression chords were joined together could be raised or lowered with the use of jacks; this would ensure the anchor arm was kept at the correct design elevation during its construction. Eventually the anchor arm would work itself free of the falsework as weight was added to the cantilever section of the bridge. The falsework could then be removed for use on the north side of the river. The cantilever arm would be erected at the end of the anchor arm, one frame at a time. No falsework was needed as it would be supported as a cantilever by the anchor arm. One half of the suspended, or center span would then be added to the cantilever arm in a similar manner, by cantilevering it out from the end of the cantilever arm, one frame at a time; temporary connecting links would be installed between the cantilever arm and the suspended span to allow the suspended span to act as a cantilever during erection. Once the south and north spans of the bridge from each shore joined together in the center, the suspended span would become a simple truss, supported at each end by the two cantilever arms.

Deans, the Phoenix chief engineer who had overall managerial responsibility for the work, had decided that all of the planning and any decisions relating to erection would be made by his engineers in Phoenixville, not at the bridge site. He described his scheme for controlling the work: "To eliminate as far as possible the necessity of the erection foreman using his judgment in connection with erection, a programme was made out in the office a year or so in advance of the actual work being done, by the erection department in Phoenixville, working in conjunction with the engineering department, fixing in every detail, every operation of the traveller and the hoisting apparatus, and defining to the minutest detail how the attachments should be placed and attached, and how the material should be loaded on the cars at the storage yards. All of this programme was indicated in clear terms in blueprints, furnished to the general foreman, assistant foremen and engineers. These instructions covered every operation and included every member on the bridge."[1]

According to the Quebec Bridge Company's erection inspector, Kinloch, "You only had to follow the blueprints and the thing would raise itself." But things do not always go as planned; whereas on other jobs, considerable discretion was given to the site personnel to deal with problems that inevitably come up in the field, this was not the case for the Quebec Bridge. Kinloch noted, "The control of this work differed from that of any other upon which I have been employed in this respect, that every question between the inspectors and the contractors was referred to New York and Phoenixville for settlement, whereas in my previous experience the power to settle most questions was vested either in the inspectors or in a resident engineer who was always on the work." The field personnel could not act on their own, even in emergencies; the erection team and the inspectors could do nothing without first consulting with the engineers, who were in Phoenixville and New York, over five hundred miles away. There was one telephone line at the site, in the Phoenix office, but the line was not private and the service was spotty at best. The wire service was plagued by strikes and the mail took at least a day and a half, each way.[2]

A.B. Milliken was the top field supervisor for Phoenix, reporting directly to the chief engineer, Deans. A man with considerable experience, he'd been with the company for eighteen years and was responsible for all of the company's projects in the U.S. and Canada, dividing his time between Boston, Tennessee, Pennsylvania, and Quebec. Prior to the Quebec Bridge, the longest cantilever bridge he'd been involved with was a single-track bridge with a span of 660 feet. His involvement in the Quebec Bridge project was peripheral at best, leaving it up to the engineers in Phoenixville to deal with important issues, as instructed by Deans.

The full-time supervisor in charge of erection at the site was Ben Yenser, a thirty-six-year-old bridge man who had started with the company when he was sixteen. He was considered Phoenix's best man and had been a general foreman with them for the past ten years. Deans described him as "showing unusual qualities as an erector, being extremely careful and conscientious, and having had large experience in the handling of men." Yenser had built bridges all over the U.S., but this was his first job in Canada. Kinloch, the Quebec Bridge inspector, described Yenser as "a hustler, and like every other erector, liked to get up as many tons of steel in a month as he could." Birks, the Phoenix inspector, worked well with Yenser and made sure that Phoenixville was kept up to date and involved in all decisions regarding erection of the structure. Milliken told Kinloch that the inspection performed by Birks was provided so that Phoenix might "get the full advantage of Yenser's energies, without anything being done contrary to the wishes of its engineering department." Birks made sure the Phoenixville office was aware of everything that went on at site. Yenser understood the limits of his authority, as did his four assistant foremen.[3]

Phoenix brought with them a small complement of skilled and experienced workmen from the U.S., but the majority of the bridge workers were Canadians, many from nearby Lévis, Quebec City, and surrounding communities such as St. Romuald. A large contingent of skilled workers came from Kahnawake, a First Nations reserve on the St. Lawrence River near the Lachine rapids, eight miles from Montreal. Pronounced "Gah'nawa'geh," the name comes from the Mohawk word meaning "place of the rapids."

The band had its origin in the latter half of the seventeenth century, when French Jesuit missionaries converted somewhere between fifty and a hundred Iroquois families in a dozen longhouse villages in what is now western and northern New York. The priests persuaded the natives to go to Quebec and settle in a mission outpost, and the converts began arriving there in 1668. Among them were members of all of the tribes in the Iroquois Confederacy, the Oneidas, Onondagas,

Cayugas, and Senecas, but the vast majority of them were Mohawks. So it was that Mohawk customs and the Mohawk dialect of Iroquois became the customs and speech of the whole group. In 1676, accompanied by two Jesuit priests, they left the outpost and travelled upriver. They moved the village three times, always farther upriver, until they finally settled at the foot of the Lachine rapids in 1719. They named it Kahnawake. Following the Treaty of Paris in 1763 in which the French gave up their lands in North America, the British conquerors granted the Mohawks title to the reserve, under the supervision of the Indian Department.

The Jesuits had tried to get the Mohawk Indians to become farmers, but the men refused to give up their nomadic ways; they cleared the land, but the farming was done by the women. Many men from the first generation born on the reserve joined the fur trade, carrying trade goods in the great fleets of canoes that travelled to remote places on the St. Lawrence and its tributaries, and returning with bales of fur. Up until the 1830s, practically every youth from the reserve worked in freight canoes as soon as they turned seventeen. As the fur trade declined, the Kahnawake men became involved in the St. Lawrence timber-rafting industry. They were famous for their skill in running immense rafts of oak and pine over the Lachine Rapids. Then, in 1886, the railroad came to their community. Life for the Kahnawake was about to change.

The Canadian Pacific Railway Company planned to build a cantilever bridge across the St. Lawrence River at Lachine and needed land from the Kahnawake reserve for their bridge abutment on the south shore. In return for the use of the land, the CPR and the builder, the Dominion Bridge Company, agreed to hire the local natives to work as ordinary day labourers, building the stone piers for the bridge and unloading materials near the site. According to Dominion Bridge records, the Kahnawake were not happy with this arrangement, "They would come out on the bridge itself every chance they got. It was quite impossible to keep them off. As the work progressed, it became apparent to all concerned that these Indians were very odd in that they did not have any fear of heights. If not watched, they would climb up into the spans and walk around up there as cool and collected as the toughest of our riveters, most of whom at that period were old sailing-ship men especially picked for their experience in working aloft. These Indians were as agile as goats. They would walk a narrow beam high up in the air with nothing below them but the river, which is rough there and ugly to look down on, and it wouldn't mean any more to them than walking on solid ground."[4]

In the erection of steel structures during the late 1800s and up until the 1950s, there were three main divisions of workers: raising gangs, fitting up gangs, and riveting gangs. The steel would come to a job already cut and built up into various members such as columns, beams, chords, etc. Each piece had a series or groups of holes bored through it to receive bolts and rivets, and each piece was numbered with chalk or paint to indicate where it should go in the structure. Using a crane or a derrick, the men in the raising gang hoisted the pieces up, set them in position, and joined them by running temporary bolts through a few of the holes. The men in the fitting-up gang would then use guy wires and turnbuckles to make sure the pieces were positioned correctly. Then more temporary bolts were added and the joint tightened up, ready for the riveting gang. One raising gang and one fitting-up gang could keep several riveting gangs busy.[5]

Prior to the 1950s, almost all structural steel connections were made using heated rivets. These have now been replaced by high-strength bolts or welding. The reason for the change is cost. Whereas two workers can install and tighten high-strength bolts, and only one or two are needed to weld a joint, it takes a minimum of four highly skilled riveters to install rivets. Riveters earned the best wages on site, and the Kahnawake men couldn't resist. According to Dominion Bridge, "They seemed immune to the noise of the riveting, which goes right through you and is enough in itself to make newcomers to construction feel sick and dizzy. They were inquisitive about the riveting and were continually bothering our foremen by requesting that they be allowed to take a crack at it. This happens to be the most dangerous work in all of construction, and the highest paid. Men who want to do it are rare and men who can do it are even rarer, and in good construction years there are sometimes not enough of them to go around. We decided it would be mutually advantageous to see what these Indians could do, so we picked out some and gave them a little training, and it turned out that putting riveting tools in their hands was like putting ham with eggs. In other words, they were natural-born bridge men. Our records don't show how many we trained on this bridge but there is a tradition in the company that we trained twelve; enough to form three riveting gangs."[6]

There were four men in a riveting gang: a heater, a sticker-in, a bucker-up, and a riveter. The heater would lay some wooden planks across a couple of beams to make a platform for his portable, coal-burning forge, in which he heated the rivets. He had to get them glowing red-hot, so that they behaved like plastic and could be easily deformed. The three others would use ropes to hang a plank scaffold from the steel on which they were going to work. They would then climb down with their tools and take their positions on the scaffold. The sticker-in and the bucker-up would be on one side, and the riveter would stand or kneel on the other. With his large tongs, the heater would pick up a red-hot rivet from the coals in his forge and toss it to the sticker-in, who caught it in a cone-shaped metal can. At this stage, the rivet is shaped like a mushroom, with a buttonhead and a stem. Meanwhile, the bucker-up would unscrew and pull out one of the temporary bolts joining the two pieces of steel, leaving the hole empty. The sticker-in would then pick the rivet out of his can with his smaller tongs, stick it in the hole, and push it in until the buttonhead was flush with the steel on his side. The stem of the rivet protruded from the other end; the riveter's side. The sticker-in stepped out of the way and the bucker-up fit a tool called a bucking bar or dolly bar over the buttonhead and held it there, bracing the rivet. The riveter then pressed the cupped head of his pneumatic hammer, or riveting gun, against the protruding stem of the red-hot rivet and turned on the air to the rivet gun. The energy from the hammer in the rivet gun would drive the rivet against the bucking bar, compressing the tail of the rivet, making it mushroom tightly against the joint in its final buttonhead or domed shape. Once the rivet cooled, it contracted, tightening the joint even further. This operation was repeated until every hole that could be got at from the scaffold was riveted up, and then the scaffold was moved.

The heater's platform stayed in one place until all the work within a rivet-tossing radius was completed, usually thirty or forty feet, depending on the heater's skill at tossing the rivets. The men on the scaffold knew each other's jobs and were interchangeable. The riveter's job was bone-shaking and nerve-racking, and so the other men would take turns swapping with him to give him a rest.[7]

Immediately following the completion of the bridge at Kahnawake, the Dominion Bridge Company began work on the jackknife Soo Bridge, which connected the twin cities of Sault Ste. Marie, Ontario, and Sault Ste. Marie, Michigan. For the Kahnawake, being away from home for long stretches of time was something they were accustomed to; their culture and nomadic way of life was a perfect fit for the construction industry. The three Kahnawake riveting gangs travelled the 600 miles to the Soo, and each gang took along an apprentice from the reserve. During the two years it took to build the bridge, the Kahnawake Indians turned the job into a college for riveters. As soon as one apprentice was trained, they'd send back to the reservation for another one. Eventually, there'd be enough men for a new riveting gang. When the new gang was organized, there'd be a shuffle-up. A couple of men from the old gangs would go into the new gang and a couple of the new men would go into the old gangs—the old would balance the new. This practice continued on subsequent jobs, and by 1905 there were over seventy skilled bridge men in the Kahnawake band. Over fifty of these men signed up to work on the Quebec Bridge as members of Local 87 of the International Association of Bridge and Structural Iron Workers' Union. Referred to at the time as "Caughnawaga Mohawk Indians," these workers were considered to be among the best steelworkers in North America.[8]

CHAPTER 9

THE DROPPED CHORD A9-L

AFTER THE CONTRACT WAS SIGNED, the Phoenix Bridge Company's parent, the Phoenix Iron Company, expanded its facilities and acquired new and larger machinery and tools to fabricate the structure. Steel fabrication began in Phoenixville in June 1904. Norris was the manager of the Phoenix Iron Works shops and before the fabrication started, he placed notices around the shops, calling the attention of his men to the fact that they were about to start work on what was going be the largest bridge in the world, and he asked every man to do his best. Norris himself never took a vacation from 1903 onwards and worked in the shop practically every day, except for five days in October 1905 when he went to the bridge site.

Just as fabrication was getting underway, Phoenix built a full-scale wooden model of one of the anchor arm bottom chord splices, showing the two chords coming together, complete with the four ribs and splice plates, right down to the rivet heads, to illustrate to the shop workers the complexity of what they were building and the importance of building it right. Later that fall, they spliced the completed steel chords 1 and 2 of the anchor arm together under the shipping crane in their yard, and asked Cooper to come to Phoenixville to see how well the work went together. Cooper travelled to Phoenixville a week later and was evidently pleased. During the last week of November, they spliced chords 4 and 5 of the anchor arm together, finishing late on a Saturday night. Over the next couple of weeks, they made several attempts to get Cooper to pay them another visit to see the chords, but he refused. They were surprised as they had been told many times by the two Phoenix senior engineers, Deans and Szlapka, that these cambered chords were probably the most important sections in the bridge; they had to be practically flawless in order to keep the bridge in perfect alignment. After these initial tests in the shop, the size of the structure and of the individual members made it practically impossible for Phoenix to follow its normal practice of test-fitting all the pieces together as they came out of the shop, prior to shipping. There just wasn't enough room, or time.[1]

All of the structural steel members, including the bottom compression chords, arrived at site assembled as individual units, with as much work as possible having been completed in the shops in Phoenixville. Some members weighed as much as ninety-eight tons and measured over 100 feet long. For the southern span, the pieces were delivered by rail to the storage yard at Chaudière, where they were unloaded and stored until needed. They would then be loaded onto rail cars and taken to the bridge site a mile away. The Belair yard on the north shore was similarly set up for work on the north span. The two yards were sixty-seven feet wide and 1,000 feet long and each yard was served by two fifty-five-ton electric gantry cranes, running on rails the full length of the yard. The yards could hold up to 12,000 tons of material each and there was enough room to assemble the eyebars into complete panels ready for erection.[2]

A considerable amount of material was delivered to the Chaudière yard during the winter of 1904 but erection at the bridge site couldn't begin until the rail line between the Chaudière yard and the bridge site was completed in July 1905. Horace Clark was the Phoenix foreman in the yard; he received and unloaded the metal, checked it, and stored it until it was needed at the bridge site. The pieces were not re-measured in the storage yard, just recorded and matched to the shipping documents. Clark and his men rarely had trouble handling the heavy pieces, except for one bottom chord section, a mishap that would prove fateful.[3]

In April 1905, Clark and his men were working in the Chaudière yard, handling chord A9-L, one of the bottom compression chords of the anchor arm. The designation meant that this was an anchor arm member, in the ninth panel of the truss, on the left, or west side of the structure. This particular chord had previously been damaged prior to leaving Phoenixville; a chain had been used improperly to lift the chord following its fabrication in the shop, resulting in a

THE DROPPED CHORD A9-L

half-inch bend in one of the chord's inside ribs. The bend was readily apparent as it rendered even narrower the already small gap between the two center ribs; the mark left by the chain was also plainly visible. The shop had been unable to take the bend out without heating or cutting the rib apart, and so it was allowed to remain—the Phoenixville engineers had examined it and pronounced it safe.[4]

Clark was carefully following the blueprint rigging instructions from Phoenixville for this eighty-seven ton, fifty-foot-long chord. The chord was being lifted by both cranes with the use of specially designed sixty-ton hooks, one on each end. The chord was about five feet off the ground when a connecting link broke on one of the hooks. The free end came crashing down, striking a steel plate lying on the ground. The hook on the opposite end came loose, and the other end of the chord came down on a pile of eyebars.

Chord A9-L was seriously damaged; the splice plates and bottom cover plate at one end of the chord were destroyed and the flange angles on two of the ribs were sheared off at the bottom. The ribs themselves were slightly bent and some of the diagonal lattice angles attached to the four ribs were damaged; the upright legs on the lattices had bent down onto the other leg. This second accident was much worse than the one in the Phoenixville shop. The engineers in Phoenixville were called upon to examine the chord and decide whether or not it could be salvaged.

The Phoenix chief design engineer, Szlapka, travelled to site with C.W. Hudson, Birks's predecessor at site. Hudson had been at site during the early stages of erection of the approach spans and had also designed the big traveller crane that was being used for erection on the bridge. The two engineers carefully examined the damaged chord and determined that it could be salvaged. They returned to Phoenixville to prepare specifications for the repairs. The specifications were received on July 8th and the chord was repaired in the yard, under Clark's supervision. Kinloch, who'd arrived at site on July 1st, saw the chord before it was repaired and then witnessed and inspected the repairs. The broken flange angles were cut off far enough back so that they were square, and the joints where the angles touched were clipped and filed, to Kinloch's satisfaction. Clark tried heating another damaged angle with a kerosene Wells light, but he couldn't get it hot enough, so he used a ram to straighten it. New splice plates were fabricated and installed, and the bent lattices were cut off and new lattices installed. The bends in the ribs themselves were less than half an inch, which the engineers considered acceptable, and so they remained. The ribs of chord A9-L now had several bends in them, but they were all within half an inch. The new splice plates that were added would also require more field riveting since one-half of the old plates that had been damaged had previously been riveted to the ribs at the shops in Phoenixville.[5]

The Quebec Bridge Company's chief engineer, Hoare, accompanied Kinloch on one of his inspections of the chord and looked over the repairs. Hoare, Kinloch, Clark and Szlapka examined the repairs and all appeared satisfied. Phoenix's resident engineer of erection, Birks, arrived at site after the chord was installed. He hadn't seen the damage that had been done to the chord, nor did he witness the repairs. He would nevertheless have very strong opinions about the dropped chord and the state it was in when it went into the structure—opinions that would prove fatal.[6]

The load carried by a bottom compression chord could only be transferred to the adjoining chord if the machined ends or edges of the four ribs were flush with the ribs of the adjoining chord. However, the designers intended for this to happen only when the bridge was fully loaded. For this reason, the bottom chords were cambered during erection; camber is the amount of curving or arching used to counteract the effects of a load. The lower compression chords of the Quebec Bridge were erected in such a way that, initially, the edges of the four ribs of the adjoining chords made contact only at the top or at the bottom of the ribs; as more load was added during the erection of the structure, the chords would gradually settle, bringing the edges of the ribs flush together over their entire fifty-four-inch depth.

This method of erection also meant that initially, the holes in the splice plates that were used to join the four individual ribs of the chord would not line up perfectly with the holes in the splice plates of the four ribs of the adjoining chord. There would be some overlap between the holes but they wouldn't be aligned enough for the erecting crews to fill the holes with the specified one-inch diameter bolts to join the members together prior to riveting. In order to join these chords during erection, smaller bolts of varying sizes would be used, some as small as five-eighths of an inch in diameter. These smaller bolts would be replaced with larger bolts as the camber decreased and the holes lined up. As weight was added, the joints would eventually close and the holes would line up. The bolts would then be removed, a few at a time, and replaced with permanent, heated rivets.

Even once a joint was closed, it was not uncommon for the holes not to line up. This could be from a number of factors, including poor fabrication in the shop. If it became impossible to get the holes to line up enough to drive the rivets, the holes had to be reamed out to make them bigger. In this way, the joint would be fastened but the members themselves would not be joined together as intended.

As the ribs of the adjoining chords came together initially, not only would they be opened at the top and in contact at the bottom, or vice versa, they might also be out of line with each other horizontally. Four feet from the joint between the ribs, the spacing between the upper half of the ribs was maintained by means of diaphragm plates between the ribs that had been installed and riveted in the shop in Phoenixville. However, these plates did not extend the full fifty-four-inch depth of the ribs, and so all the imperfect horizontal matchings occurred at the bottom of the chords. Where the ribs were out of line at the bottom, they would have to be pulled or pushed into place with jacks or mauls and steel wedges.[7]

The size of the structure, compounded by the curvature and resulting angles introduced by the unique design, made joining the lower compression chords the greatest challenge faced by the erectors. The Phoenix designers, as well as Cooper, viewed these bottom chords as the weakest part of the structure, requiring the closest attention. But the bottom compression chords were sadly neglected, compared to the attention paid to the top chord tension eyebars.[8]

Numerous tests were conducted on the eyebars, which had been designed by Cooper himself. At the end of 1904, Cooper was made aware by Edwards, the Quebec Bridge Company's inspector in Phoenixville, that a number of eyebars had elongated when tested to the intended strains. Cooper thought the matter serious and suggested that Phoenix should enlist the co-operation of other bridge companies in making a thorough examination of the problem. It appeared to Cooper as though Phoenix was more desirous of hiding the matter rather than exploiting it—they asked Cooper not to make the matter public. He reacted by sending a letter to Edwards on January 8, 1905, directing him to accept no more eyebars for the Quebec Bridge Company until further orders. Edwards was also to give a copy of the letter to Phoenix. Phoenix immediately carried out the investigation to Cooper's satisfaction. For the first time, Cooper had stopped part of the work in the shops through the Quebec Bridge Company's inspector. The eyebars would prove to be the strongest part of the structure. The weakest part of the structure, the bottom compression chords, were never subjected to any testing.

Although there was no machine in existence large enough to perform such testing on members as large as these, it would have been possible to build smaller scale sections that could be tested, but this wasn't done until it was too late.

CHAPTER 10

THE 1905 & 1906 SEASONS

ASIDE FROM ITS OWN WEIGHT, the bridge also had to carry the weight of the equipment used for its erection. A moving gantry crane, or "traveller," was designed specifically for this bridge. This steel structure weighed over a thousand tons, measured fifty-four-feet long, 105-feet wide, and was 220-feet high. It was designed to extend twenty feet on either side of the main trusses, allowing for the completion of the full structure, including the external sidewalks, as it advanced from one panel to the next. This large, or main, traveller was fitted with four 125-horsepower electric hoists that could handle pieces weighing up to 105 tons.[1]

Power for the project was supplied by the Canadian Electric Company from its hydro station at the nearby Chaudière Falls. Electricity was used to run the hoists on the two travellers and the compressors for riveting, reaming, and drilling, eliminating the use of fire on the structure, with the exception of the rivet-heating forges. Electric power was used more extensively on this structure than in any previous work done by Phoenix; they were convinced of its superiority over steam.

The main traveller was to be used for all erection, except for the suspended span near the center of the bridge, where a small traveller would be used. There were two reasons for this: firstly, Cooper wanted them to use the small traveller, since it was only a quarter the size and weight of the large one. This would result in a considerable saving of metal in the suspended span as the weight of the traveller during its erection would be the heaviest load the span would ever have to carry. The second reason was political. As early as 1903, the Dominion Government had expressed a desire for the bridge to be completed in time for the 1908 Tercentenary of Quebec City, when it could be officially opened by the Prince of Wales, the future king. Phoenix planned therefore, to use the small traveller to erect the suspended span of the south arm, thus allowing the large traveller to be moved to the north side of the river in the fall of 1907. This would accelerate the work and potentially allow the bridge to be completed in 1908.[2]

Although Phoenix was ready to start erection of the steelwork in the spring of 1905, it didn't begin until July 22nd. The Quebec Bridge Company was responsible for providing the rail connection between the Chaudière yard and the bridge site, but it wasn't ready until July 9th. Prior to this, all of the metal for the anchorages and approach spans, as well as the materials for the falsework and traveller, had to be sent to Lévis or Quebec City and then taken to the bridge site on barges. Aside from the delays to the work, Phoenix also experienced considerable difficulty with congestion of materials in their yard in Phoenixville and at Belair on the north shore. Since the bridge pieces for each side of the river were fabricated at the same time to better match the templates, a large amount of material was being stored. Tensions eased when the rail connection was completed. Steel erection finally got underway and the stockpiled material started going into the structure.

By September 1905, the bottom chords of the anchor arm were in place, including chord A9-L. It looked bad, but it was safe, according to the engineers. The Quebec Bridge Company's inspector, Kinloch, noticed bends in the ribs of three other lower chords, after they were set in place, but before any stress came on; they didn't look straight and appeared wavy; as much as a half-inch. Kinloch discussed the bends with the other inspectors, McLure and Birks, but they all decided they were minor shop bends and thus not important. Hoare and Yenser, the Phoenix general foreman, were also aware of the bends but they, too, looked upon them as minor defects, which would not in any way affect the strength or stability of the bridge.

The 1906 construction season began in April and was proceeding well until June, when Cooper stopped a portion of the work. The erection crews were working on one of the structure's centre posts above the main pier. McLure had sent the consulting engineer a sketch showing poor shop workmanship on one of the post's bearings. Cooper immediately wired Hoare, the chief engineer for the Quebec Bridge company; "Do not allow post CP-1 erected until top is made level. Notify McLure." Hoare immediately issued instructions to the Phoenix foreman to this effect, and the work on the post was stopped. The Phoenix Bridge Company immediately sent one of its engineers from Phoenixville to check McLure's measurements. The shop defects were confirmed, and the post made good, based on detailed instructions from Cooper. The work of the contractor had been stopped by Cooper, through Hoare. The work on the post did not resume until Cooper was satisfied that the defect had been corrected.

The ten panels of the south anchor arm were all in place by the end of June 1906. In July, McLure wrote to Edwards, the shop inspector for the Quebec Bridge company in Phoenixville, expressing concerns about the bends in the ribs of the compression chords. "On a number of the compression chords that we have erected, in sighting end to end, the webs in places are decidedly crooked, and show up in wavy lines apparently held that way by the lacing angles. This makes a very bad appearance, for a person seeing a member like that, and knowing it to be in compression, would at once infer that it had been over-strained sufficiently to bulge the webs. As to its actual effect in a number of cases, I have figured out there is no possibility of this causing trouble, as long as the lacing in the members in question is intact." Late in September, McLure reported these deflections to Cooper. Although he didn't like the distortions, Cooper didn't see that anything could be done about them at that stage, and so nothing was done—the work carried on. The bridge workers saw the bends and wondered whether it was a sign that the structure was overstressed.

The south anchor arm remained on the falsework between the two piers until August 1906. The cantilever arm was well underway by that time with eight panels extending 450 feet out over the water. Phoenix issued detailed instructions for the removal of the falsework, such that the anchor arm sections nearest to the main pier would lift first. This would enable the same falsework to be erected on the other side of the river, as planned. However, Phoenixville decided to change the sequencing and issued revised written instructions to this effect to its superintendent, Milliken. Due to an oversight, the Quebec Bridge Company inspector, McLure was not advised of the change. When he saw the men removing the falsework at what he considered to be the wrong end, he immediately sent a letter to Milliken complaining about the lack of notification and practically demanded that the work be stopped. Tempers flared and Milliken notified Deans, who criticized McLure for his "Lack of experience." The controversy was eventually settled by a letter from Hoare to Deans, explaining the need to keep Cooper and himself informed of all important matters, through their full-time on-site representative, McLure. In his letter, Hoare made reference to Deans's comment regarding McLure's lack of experience and told him it was uncalled for. Deans sent McLure a letter of apology and invited him to raise any future concerns with Milliken.[3]

At their next meeting in New York, Cooper and the Phoenix chief designer, Szlapka, discussed the friction that had developed between McLure and Milliken. In Cooper's view, Phoenix, "Did not recognize the rights of anybody but themselves to control the erection." Cooper told Szlapka that, "The Phoenix Bridge Company were not the only parties who had a financial interest in this structure, that the parties whom I represented, the Quebec Bridge Company, had paid for the structure as it stood, that it belonged to them, and they had an interest in seeing that it was not risked or injured, and while I always endeavored to get along amicably with everybody, if it came to a point of determining my right or the right of any employee under me to protect the property of the Company, I thought they would find themselves in the wrong." Cooper made it clear to the chief design engineer that he was the boss—he had stopped the work before and would do so again if circumstances demanded it.

By November 26th, erection of the south cantilever arm was nearly completed. A total of 21 million pounds of steel had been erected that year on the southern span. Work on the North Shore of the St. Lawrence had started in July, with the erection of falsework for the north anchor arm.

By the end of November, all but eight of the forty lower chord joints were closed. The camber openings were gone and the ribs of the chords were flush against each other, ready for riveting. The joints were not to be riveted up, however, until the following year, once the suspended span was complete, when the members would be under full load. The instructions from Phoenixville regarding the joints between the chords were clear; once the joints were closed and prior to the riveting of the splices between the chords, the smaller bolts that had been used to erect the members were to be replaced with larger bolts and all of the bolts were to be tightened.

The two outside ribs of the lower chords were clearly visible and could be easily inspected. This meant that the bolts in these outside ribs were sized right and continuously tightened. However, the two inside ribs of the chords were hidden from view by the top and bottom cover plates that joined the ribs together. The inside ribs would not be visible or accessible until the bottom cover plate was removed just prior to riveting. This meant that the joints between the inside ribs of the adjoining chords were loose. The smaller bolts that were used to join the inside ribs together were not replaced with larger bolts, nor were any bolts systematically tightened up as they worked loose with the adjustment of the structure. The joints between the inside ribs remained that way until the following year. This resulted in the ribs deforming where they came together, and this would not come to light until June 1907, when the bottom cover plates were removed.[4]

CHAPTER 11

1907 EARLY WARNING SIGNS

B
Y THE END OF THE 1906 construction season, the southern span struck a magnificent pose. The 500-foot-long anchor arm between the two piers near the shore seemed to carry the 562-1/2-foot-long cantilever arm with ease, as it extended from the main pier out over the St. Lawrence. The gentle curves of the top and bottom chords of the trusses added elegance to what would become the longest clear span bridge that had ever been built.

The Phoenix superintendent's plan for the 1907 construction season was to finish the erection of the south half of the suspended span, get all of the riveting done on the cantilever and anchor arms, and remove the main traveller and erect it on the north side of the river. He planned to complete the north shore false work, ready for erection of the anchor arm the following spring, in 1908. Phoenix had verbally committed to do its best to complete the bridge in time for the 300th anniversary, but Milliken knew that the work wouldn't be done until 1909. Phoenix had been paid just over $3,000,000 as of March 31st.

Work began on the north side of the river in March, getting the Belair yard ready to receive more material from Phoenixville. On the south shore, erection crews got underway installing the connecting links between the cantilever arm and the suspended span with the big traveller. Once the connecting links were in place, the big traveller was to be dismantled and moved to the north shore. The small traveller would then be used to complete the erection of the suspended span.

Close to 300,000 field rivets were required to join the pieces of the south half of the bridge together, but prior to 1907 only about 50,000 rivets had been driven. Szlapka and Cooper had planned to delay the riveting of the structure until the erection of the south half of the bridge was completed and all joints had their full stress. On May 10th, however, they decided that riveting could be carried out at once for all joints where the connecting pieces had full contact with each other.

There had been only one or two riveting gangs at the site prior to 1907, but by June, there were seven or eight four-gangs riveting up the joints between the compression chords of the anchor and cantilever arms. When riveting the joints between the chords, the men would begin by hanging a scaffold under the joint to give them safe access. They would then remove the bottom cover plate and set it down on the scaffold. Kinloch had told them to always install two small angles to temporarily take the place of the bottom cover plate to keep the ribs properly spaced apart. When the four-gangs removed the bottom cover plates in the anchor arm, they saw the result of the inside rib splices not having been properly bolted up and continually tightened.

McLure described what they found in a report to Cooper on June 15, 1907. "In riveting the bottom chord splices of the south anchor arm, we have had some trouble on account of the faced ends of the two middle ribs not matching; the lower ends of the middle ribs of the abutting chords were out of line by 1/8 to 1/4 inch, this offset decreasing to nothing near the mid-depth of the ribs. This has occurred in four instances so far, and by using two 75-ton jacks we have been able to partly straighten out these splices, but not altogether. These were probably in this condition when erected but owing to the presence of the bottom cover plate, it was then impossible to detect them, and it was only when this plate was removed for riveting that the inequality was noticed. The chords found in this condition were between A3-R and A4-R, A7-R and A8-R and A8-R and A9-R, in the east truss and A8-L and A9-L in the west truss. You will note that this occurs only on the inside ribs, which are provided with but a single thin splice plate each. I think that a heavy plate on each side of these ribs, bolted up tight when chords were erected, would have remedied this, i.e., drawn the ribs together till the faced ends matched."[1]

At this stage, McLure thought that the adjoining ribs must have been bent that way when they went into the structure. In his view, the misalignments couldn't possibly be due to the chords being overstressed, since there was a great deal more steel to be added to the structure; the chords weren't yet under their full design load. McLure was the first person to blame the shops in Phoenixville for the misalignments and bends in the chords.[2]

When Cooper read McLure's report, he wasn't concerned. He replied on June 17th, "Make as good work of it as you can. It is not serious. It would be well to draw attention to as much care as possible in future work to get the best results in matching all the members before the full strains are brought upon them." Cooper's reasoning was the same as McLure's; the members didn't yet have the full strains upon them and so the bending couldn't possibly be due to their being overstressed. He didn't respond to McLure's suggestion of using two heavy splice plates to join the inside ribs together, similar to the design for the two outside ribs. The work continued, without modification, and the four-gangs did their best to straighten the edges of the adjoining ribs, prior to riveting the single splice plates to the inside ribs.

By early August, the suspended span extended well out from the cantilever arm, adding still more load to the completed sections. A four-gang working on the lower compression chords of the cantilever arm removed the bottom cover plate between chords C7-L and C8-L in the west truss, and once again found the edges of the two inside ribs out of alignment, but here the ribs were three-quarters of an inch out of line—three times as bad as those previously found. McLure reported the discrepancy to Cooper on August 6th, pointing out that owing to the limited space between the two inside ribs, it was impossible to jack these ribs into place. He suggested that a diaphragm plate be installed between the two inside ribs near the bottom to hold the splice against the increased thrust the ribs would have to bear due to their being out of line. Birks, the Phoenix resident engineer, agreed with the proposal and sent a similar report to Phoenixville.[3]

This time Cooper was concerned. He sent Deans a telegram on Thursday, August 8th. "Method proposed by Quebec for splicing joints at lower 7 and 8 chords is not satisfactory." He asked, "How did bend occur in both chords?"[4]

Deans replied the next day, advising Cooper that his chief design engineer, Szlapka, had been at the bridge site the previous day, and that he expected him back on Saturday "with full information concerning chord joint; will then write to you fully."

Cooper sent Deans a note with his own ideas for correcting the misalignment, and asked him to let him know what he and Szlapka proposed to do; he added, "It is a mystery to me how both these webs happened to be bent at one point and why it was not discovered sooner."[5]

There was a strike at site on Thursday, August 8th—the dispute was over travel costs being deducted from the workers' wages. The previous month, Phoenix had hired twenty-two bridgemen out of New York, but only ten reported for duty and several of these left within days of their arrival. Phoenix hired another fifteen from Buffalo, but a number of those also left within a few days. Phoenix had paid the expenses for these men to travel to Quebec; if a worker left soon after arriving, the company deducted the travel costs from any earned wages. The strike lasted only three days, but many men went back to the U.S. and told their union brothers about the Phoenix practice of deducting travel costs. As a result, it became more difficult to attract workers to the site.[6]

The president of the union, D.B. Haley, had started on the job in June; he'd been assigned to work on the small traveller. On the first day of the strike, with nothing to do, he decided to have a look at the bridge members that some of the men had been talking about. He went down with two of his co-workers, Joe Ward and George Cook, the secretary of the union, to see the joint between chords C8-R and C9-R on the cantilever arm. The two thin splice plates used to join the inside ribs together were lying on a scaffold, along with the bottom cover plate. Haley climbed down onto the scaffold and looked up inside the chord; the faced edges near the bottom of the inside ribs were not flush together. Haley had a ruler with him; he measured the misalignment; it was about half an inch. There were four hydraulic jacks being used to try to get the rib edges to line up flush so that the plates could be riveted on. Haley had been in the business for nine years, but he'd never seen anything like this. It seemed pretty clear to him that too much weight was being put on before the joint was riveted up.[7]

On Saturday, three days after the strike began, a union meeting was held, and a vote was taken. By this time, many of the discontented workers had left the site, and the majority voted in favor of returning to work under the original agreement. Most of the men who had voted against returning to work also left, reducing even further the workforce on both sides of the river. Although the strike ended on Saturday, work didn't get underway until the following Tuesday. The safety of the structure had never come up during any of the discussions between the union and the Phoenix managers.

Szlapka didn't return to work in the Phoenixville office that Saturday as planned, so Deans telegraphed Cooper that they would report to him the following Monday. On Monday, Szlapka told Deans that he hadn't had time to examine the splice between chords C7-L and C8-L. In fact, during his three visits to the bridge site, Szlapka never looked at any of the lower chords and splices, the same compression chords and splices that he, Deans, and Cooper all viewed as being the weakest part of the structure. Nevertheless, Deans wrote to Cooper and told him that the ribs of chord C7-L had complete and full bearing on the ribs of chord C8-L and that only one of the ribs was bent. Deans concluded his note with, "The bend was no doubt put in the rib in the shop." The Phoenix chief engineer had no grounds for any of this; Szlapka hadn't seen the joint, nor had Deans checked with Morris, the head of fabrication in Phoenixville, as to whether the bend had in fact been made in the shop. If Deans had checked, Morris would have told him that there was simply no way that such an error would have escaped detection.

Both Deans and Szlapka thought that the misalignment wasn't important and recommended against the installation of the diaphragm suggested by McLure and Birks. As far as they were concerned, the joint was ready to be riveted up as is. Cooper replied on August 13th that the information received from McLure was different from what Deans and Szlapka reported and asked the obvious question, "How was it that with only one rib being bent, there could be complete and full bearing between the ribs?" Cooper told Deans that he could take no action until the exact facts were presented to him. He then wrote to McLure, relaying Deans's information.

McLure replied immediately. "One thing I am reasonably sure of, and that is that the bend has occurred since the chord has been under stress and was not present when the chords were placed." He also reported yet another misalignment; this one was at the joint between chords C8-L and C9-L of the cantilever arm, but the ribs were only 5/16 of an inch out of line. The young engineer had changed his mind about how the inside ribs had gotten bent. In his report to Cooper, McLure stated that although the bends were not made in the shop, one of the inside ribs of the adjoining chords must have been marginally longer than it was supposed to be, causing the ribs to bend at the end, once the stress came on. Whatever solution Cooper and Szlapka came up with for the serious misalignment at joint C7-L and C8-L, McLure wrote, could be applied here as well.[8]

Cooper was unconvinced, but one thing he was certain of was that the bends could not possibly be the result of the chords being overstressed; there had to be another explanation. He replied to the young engineer, telling him that none of the explanations for the bent chords stood the test of logic. "I have evolved another theory, which is a possible if not probable one. These chords have been hit by those suspended beams used during the erection, while they were being put in place or taken down. Examine if you cannot find evidence of the blow, and also make inquiries of the men in charge." It never entered the senior engineer's mind that the structure was overstressed; how could it be, he reasoned, it wasn't yet fully loaded. While the engineers deliberated, the joint with the serious misalignment remained in its loose, unriveted state.[9]

Later that day, McLure, who by this time was getting nervous, climbed down to the lower compression chords of the anchor arm and went directly to chord A9-L, the damaged chord. He sighted along each of the four ribs, paying particular attention to the center rib that had the mark of the chain on it, to see if there were any noticeable differences. But the ribs looked to him to be pretty straight, at least to within half an inch. He was relieved to some extent, but not entirely.

McLure replied to Cooper's letter the next day and pointed out that Deans had misinterpreted the report that he and Birks had sent regarding the splice between C-7L and C8-L of the cantilever arm. He referred to the sketch that was included in the report, which clearly showed that the edges of the two inside ribs of both chords were bent where they came together. McLure told Cooper that "It would be hardly probable that these two ribs from different chord sections should be bent the same way, exactly the same amount in the shops to dimensions 1/2-inch to 3/4-inch less than called for. I am reasonably sure, as I said before, that this condition did not exist before the erection of these chords, as I have personally inspected every member yet erected in this bridge thus far, except the bottom chords of the anchor arm, on the cars just before erection, looking particularly for bends in the ribs of compression members, and wherever discovered have taken measurements of the amounts and recorded them. If these ribs in the cantilever arm had been this much out of line before erecting, it would be well-nigh impossible to miss seeing them."[10]

Alexander Beauvais was a Kahnawake Indian, with six years' experience as a steel worker. He had been hired by Phoenix in 1905 as head of a four-gang of riveters. His supervisor was C.E. (Slim) Meredith, the rivet boss. Since May, Beauvais's four-gang, and three other gangs, had been riveting the joints between the compression chords of the anchor arm. Beauvais and his four-gang had been working on the joint between chords A9-R and A10-R of the anchor arm on the

Quebec side for almost two weeks. When they removed the bottom cover plate, Beauvais looked down the fifty-foot-long chord A9-R; he could see that near the center of the chord, the two inside ribs were bending inward, towards each other. It wasn't as easy to see whether the outside ribs were bent because there was a space of fifteen and a half inches between the flanges on the inside and outside ribs of the chord. The flanges on the inside ribs were only two inches apart and so the bend was quite noticeable and easy to see. Beauvais had heard that there were also bends in the ribs of the chords in the cantilever arm, but although this bend was readily apparent, neither he nor anyone else in his four-gang raised it with anyone.

There were 280 holes altogether on the inside ribs at this joint; 140 for each rib. Although most of the holes had bolts in them, they hadn't been tightened by the erectors. Beauvais's gang had to first tighten the bolts before they could start riveting. Kinloch had told them to follow a certain procedure when riveting these joints; once the bolts were tightened, they were to remove no more than five or six at a time. They could then drive the hot rivets into these holes and proceed to take another five or six bolts out. Yenser had told Beauvais the same thing. The inside ribs were to be driven first and then the outside ribs. Once this was done, they could install the top cover plate, followed by the bottom plate, in sequence.

Once the six splice plates and top cover plate were riveted on this joint, Beauvais and his men had trouble getting the bottom cover plate back on. The ends of the inside ribs were bent, and they had to use drift pins to line up the holes. When the joint was complete, Kinloch inspected it and had two or three rivets cut out for testing. They drove in new rivets and moved on to the lateral bracing. They couldn't work on the joint between chords A9-L and A10-L on the Montreal side because the big riveting gun they needed was being used by the riveters on the cantilever arm. They completed the laterals from one side to the other, and when the gun became available, they moved to the Montreal side.

McLure got sick in August and had to spend nearly a week in hospital in Quebec City. On the morning of Tuesday, August 20th, Joe Ward, one of the bridge men who had looked at the joint on the cantilever arm with Haley and Cook earlier in August, lost his balance while he was trying to draw a pin from a shackle out on the floor of the third panel of the suspended span. He fell 170 feet into the river, but his body didn't come to the surface. All work was suspended for the rest of the

1907 EARLY WARNING SIGNS

day. This had been the fifth fatality since erection had started in 1905. According to the Phoenix chief engineer, Deans, the company was not to blame, "There were only five fatal casualties during the entire time, and each of these casualties was the result of the individual action of the man."[11]

Kinloch telephoned Hoare at 9:00 a.m. to give him the news and told him that very little work was being done that day on account of the fatality. He asked Hoare to convey the information to McLure at the hospital and tell him not to worry about getting to work that day.

Later that afternoon, Kinloch took advantage of the lull in activity and inspected some of the compression members. A boy was painting some recently installed rivets on chords C8-R and C9-R of the cantilever arm. Kinloch was coming down chord 8 and the new paint on the rivets drew his attention to the splice. He fancied he could see a curve right at the splice, and when he looked more closely, he found that indeed there was a bend. Kinloch looked at the rest of the chord and found there was also a distortion along the length of the ribs; both sides of chord 8 went out from the centre. This chord had been erected in early 1906 and he hadn't noticed these bends before. The joint had been fully riveted up and that was what worried him as much as anything, because he was positive that the ribs were straight when it was riveted up. He also looked closely at chords 9 and 10 of the cantilever arm and they too appeared somewhat distorted—all three chords, 8, 9, and 10, one after the other. He went looking for Birks and the two of them examined the three chords. Birks didn't think it was anything serious, and after talking it over with the young engineer, Kinloch thought he must have been mistaken. Birks had convinced him it wasn't anything to worry about.

The next morning, though, Kinloch still felt uneasy about what he'd seen the day before. He decided to go down to the cantilever arm again and examine chord C8-R. When he got there, Kinloch imagined that it was getting worse. One of the Kahnawake workers was there and Kinloch asked him if he had noticed anything odd about the chord. The worker told him that it didn't look to him like it had been bent like that before. Kinloch asked him if he was sure it hadn't been bent before, but the worker couldn't be certain. The bend was in the body of the chord itself, along each of the four ribs. Kinloch thought he must be nervous and that he was seeing more than what was really there, but when he looked again, he could plainly see that all four ribs were bent. What was really odd was that they were now bent in different directions; three of them were bent towards Montreal at the top, and partway down they were bent towards Quebec. The fourth rib, one of the inside ribs, was bent towards Montreal, but very slightly. He could see that the bend crossed—it was the shape of a long letter S, or a question mark. He looked at the lattices tying the four ribs together, but he didn't notice anything unusual about them. There were no distortions, which was odd as the bent ribs should have caused the lattices to be strained and distorted. He decided that he would keep a close eye on the cantilever chords to see if there was any further movement in them. After that, he climbed down three or four times a day to look at them.[12]

It was now August 21st. What Kinloch had witnessed were signs that the structure was failing; the ribs of the chords in the cantilever arm were buckling from the ever-increasing strain brought on by the addition of more steel.

On August 23rd, a problem was discovered at the splice between chords C5-R and C6-R of the cantilever arm. One of the center ribs was out of alignment at the bottom by about a half inch. This brought the number of misaligned compression chord joints to eight, out of a total of thirty-six.

Since work had started in March, Milliken had spent about half of his time at the site, but he'd been at site full-time since August 6th. High winds, a shortage of men, and the three-day strike had delayed the work. They had been forced to stop dismantling the big traveller a month earlier. On Saturday, August 24th, Milliken sent one of his people to Boston to get more men; he wanted to increase his workforce by fifteen to twenty-five. He planned to travel to Phoenixville the following day.

Milliken wasn't looking forward to his meeting with Deans on Monday to discuss the progress of the work. Prior to leaving, he met with Yenser and his field supervisors. At that point, there were two joints on each of the west and east bottom chords of the anchor arm that hadn't been riveted up. He told his men they had plenty of riveters and that he wanted these joints riveted up right away. They could then focus on finishing up the riveting on the cantilever arm. He wanted to get this work done as quickly as possible so they could go ahead and paint it—he wanted all the riveting done before painting started.[13]

The correspondence continued between the engineers regarding the bends at the faced edges of the inside ribs of the chords. Deans wrote to Cooper, clarifying his understanding of the condition of the inside ribs at the chord splice C7-L and C8-L. He confirmed that there was a perfect bearing between the four ribs of the two chords, and that, indeed, each of the two chords had one bent rib, and not one rib only, as he had previously understood. McLure reported to Cooper that he could find no evidence of the bent ribs having been hit, as Cooper had suggested. On learning this from McLure, Cooper wrote to Deans on Monday, August 26th; "Mr. McLure reports that he can find no evidence of the bent ribs having been hit and does not think they could have been struck. This only makes the mystery the deeper, for I do not see how otherwise the ribs could have been bent. When convenient I would like to discuss with Mr. Szlapka the best means of getting these ribs into safe condition to do their proper work." Deans replied the following day that he would have Szlapka call to see him in New York at the first opportunity.[14]

Eugene Lajeunesse was a bridge man from nearby St. Joseph de Lévis. On Monday, August 26th, he saw Yenser and Birks looking at a chord on the cantilever arm on the Quebec side of the bridge. After lunch, he and a few others climbed down to get a look at the chord, C9-R. The ribs were all bent in the same direction, out from the centerline of the bridge, towards the east, or Quebec side, and it was quite noticeable. Eugene thought they were bent two inches. One of the foremen, Worley, saw them and asked what they were looking at. Eugene told him: "That bottom chord is bent." Worley replied, "It was always like that." Before he could say anything else, the whistle blew. After work that day, Eugene went over to the other side of the bridge to see if the opposite chord was bent, but it looked to be all right.[15]

As the suspended span crept out over the river, the stresses on the compression members in the cantilever and anchor arms increased steadily. By this time, the splice between chords C7-L and C8-L of the cantilever arm had been sitting in its loose state for almost a month. Deans and Szlapka clearly attached no importance to the bends in the rib, and Cooper wanted to discuss it further. No one seemed to be overly concerned.

The Quebec Bridge was giving everyone ample warning- it was entering the early stages of collapse.[16]

CHAPTER 12

TUESDAY, AUGUST 27TH

D**ELPHIS LAJEUNESSE WAS A GENERAL** bridge man; he worked at everything. He and his brother, Eugene had started on July 23rd as riveters, working for the rivet boss, Slim Meredith. On the morning of Tuesday, August 27th, the whistle blew at five minutes to seven to signal the seven o'clock start. He and his brother were assigned to work at the center post on the south main pier, but before reporting there, Delphis went to look at the bent chord in the cantilever arm, number C9-R on the Quebec side. Every man he saw on the bridge that morning told him to go and see it. He walked over and looked down from the bridge deck. There was a crowd of workers staring at something. "What the hell are you looking at?" he said to no one in particular.

"Look at that chord," someone snapped back.

Delphis looked down and saw the bend in the ribs of the chord. He made his way down from the deck and stood on the chord next to his brother, who was looking at it with the other men. He turned to his brother and said, "It's dangerous."

"Oh, no," his brother joked as he jumped up and down on the chord, "It's strong enough to hold me!"

The four webs were all bent in the same direction, towards Quebec. They were bulging at least two inches, over a length of about two or three feet, just a couple of feet away from the cover plate near the splice. The top and bottom plates of the chord were on and riveted up; the work on the joint had been completed as the scaffold had been moved away. Delphis thought it was serious. He looked to his brother and said, "By God, I am going home before some accident." But he didn't leave the site; he went to work on the center post. Later that day, he went over to look at the chord on the Montreal side to see if it, too, was bent, but it looked to be all right.[1]

At nine-thirty that morning, Kinloch went down to have a look at chord A9-L on the anchor arm, the dropped chord, and when he saw it, he went numb. He'd been on that chord the previous Saturday and hadn't noticed any bend, other than the small bend that was present when it went into the bridge. This bend was quite alarming—to him at least. It was a big bend and he was satisfied right off that there was something going on. He knew that the bend could not have resulted from the accident in the storage yard. The chord had been struck in such a way that any resulting bend would have been in the opposite direction to the bend he was looking at now.

Yenser and Birks were standing at the foot of chord A10-L not far away. Kinloch called them over. Yenser looked at the chord, turned to the two young engineers and said, "That bend has never been there before. I've been over this chord too many times."

The three of them spent thirty minutes examining the chord. Yenser told them he wouldn't put up any more iron until he found out about it. Birks laughed at him and said, "You'd better wait until you find out, because when you're condemning that chord, you're condemning the whole bridge."[2]

Kinloch and Birks went up to the office to get McLure. They told him about the chord and the three of them immediately went down to measure the bend. On the way, Birks said the chord had been bent like that when it went into the bridge, but Kinloch stood his ground. He told the two engineers that "There was something seriously the matter with the chords that he didn't understand, and that they required immediate attention."[3]

The three men took measurements with a bit of fishing line stretched from the edges of the top cover plates at each end of the chord. They measured the offsets along the line at every point where the lattice angles crossed the ribs. They examined the lacing itself very carefully. They didn't notice any changes in the lacing, but they were strained awfully high. Kinloch tested the lattice angles by hitting them with his hammer, and they "sang as if they had an awfully good pull on them." The three inspectors looked at every rivet for bends or cracks, or if they were humped or sagged or bent in any way. There was no evidence of the rivets being distorted. They went to the Quebec side of the bridge to examine chord A9-R, but it looked to be all right. Then they went over to the cantilever arm and measured the number 9 chords on the Montreal and Quebec sides. The bends in the ribs of those chords were not as great as the bend in A9-L, but they were still concerned. They were at it until noon.

Kinloch and the two young engineers had their lunch at the bridge and talked it over with Yenser. They did some rough calculations of the stress that chord A9-L was being subjected to at that point in time. Birks tried to convince them that there was no danger; he told them that the bend had always been in the chord. Kinloch knew that Birks had never seen the damage that had been done to the chord, nor had Birks inspected the repairs. Kinloch had been there, and he was certain that the chord had gotten worse since it had been erected. The men didn't know what to do, and they could do nothing without involving the engineers in Phoenixville and New York. Yenser and Kinloch suggested that McLure and Birks should go to New York and Phoenixville for advice. Neither of them welcomed the suggestion, and said they'd be laughed at on arrival. They all agreed that the matter couldn't be explained properly by either phone or telegraph, and so they decided to immediately prepare reports, with sketches, and mail them to New York and Phoenixville. They also discussed whether any more work should be done pending the preparation of these reports and receipt of the responses from the senior engineers.

The four of them went up to the office to talk about whether or not they should keep erecting steel; the small traveller was ready to be moved farther out. McLure told them he thought it would be "poor policy" to either move the small traveller out or to add more weight to the structure. He argued that if anything had to be done to rectify those chords, it could be easier done at that time than after the stress had been increased. McLure and Kinloch were certain that the deflections had occurred after chord A9-L had gone into the bridge, but neither of them was sure about the bends in the chords of the cantilever arm. There was also no doubt in McLure's mind that chord A9-L could never recover itself and get back into line on its own. As an engineer, he knew that a column that was once forced out of line by direct thrust along its axis, couldn't possibly recover itself while that thrust remained, not without the application of some exterior force. Yenser was clearly worried; he told them he intended to add no more iron to the bridge until

TUESDAY, AUGUST 27TH

they found out what was causing the bending of the chords. It was finally decided that they would not move out the traveller until they got some word.[4]

Ed Britton was an electrician who worked on all parts of the bridge and the storage yards; he was always on hand when the travellers were to be moved. He'd been waiting for word from Yenser and had overheard the conversation that took place in the office. He heard Yenser say that he didn't want to move the traveller, because his life was in danger as well as others. As Yenser was leaving the office, he saw Britton standing near the door and told him the traveller would not be moved out that day. Britton had heard the men talking about the bent chords, but he hadn't paid much attention to it; they were just rumors. But now he began to think that this was serious. He went out to the small traveller and told the shop steward, Haley, as well as Cook and some others, about what he'd overheard, and about Yenser's decision not to move the traveller out. He asked them if they had heard anything about chord A9-L. Cook and Haley said they'd seen the bends in the chords of the cantilever arm but weren't aware of the bends in chord A9-L of the anchor arm. They made a plan to go down to the chord that night after work to see it for themselves.

The previous day, the storage yard foreman, Clark, had sent two temporary stringers out to the bridge; they were to be used for the erection of the suspended span once the small traveller had been moved out. Following the discussion with the engineers, Yenser ordered his men to return the stringers to the storage yard. News of Yenser's decision not to move the traveller out and to send the stringers back to the yard spread like wildfire on the job site. The men were alarmed.[5]

Yenser didn't want to use the telephone or telegraph to get in touch with Phoenixville; he knew the workmen were already anxious. Instead, he mailed a letter to Phoenixville that afternoon. "Sketch will be mailed to you today by Mr. Birks, showing buckle in bottom chord 9L anchor arm, and chords 9R and L cantilever arm, which was not noticed until today. Please advise by wire if you think this serious enough to consider the suspension of erection until more of the main traveller is removed." Birks also sent a letter to Phoenixville, in which he wrote, "We are satisfied that chord 9L anchor arm was not in this condition until recently. Its present condition was noticed for the first time today."[6]

It took McLure the rest of the day to prepare his report and the sketches showing the deflections. He wrote, "Although a number of the chords originally had ribs more or less wavy, as I have reported to you from time to time, it is only very recently that these have been in this condition, and their present shape is undoubtedly due to the stress they are now receiving. Only a little over a week ago, I measured one rib of the A9-L chord of anchor arm here shown, and it was only 3/4 of an inch out of line. Now it is 2-1/4 inches." In his cover letter to Cooper, he stated that Yenser intended to add no more iron to the bridge until he was further advised by the proper authority. McLure's report and Yenser's letter were sent that Tuesday evening; they would not arrive in New York and Phoenixville until Thursday morning, thirty-six hours later.[7]

McLure had called the Company's chief engineer, Hoare, earlier in the day and told him he needed to see him that afternoon. He didn't want to say anything over the phone as the line was not private. After sending his report to Cooper, McLure travelled to Quebec City to meet with Hoare. He explained the situation and gave him a copy of his report. He told Hoare that all four ribs of chord A9-L showed deflections towards the axis of the bridge and showed him a pencil sketch of it. The bends had been discovered that morning by Kinloch, and since these bends hadn't been noticed before, he had reported the matter to Cooper. Birks had also reported in the same manner to Phoenixville. He told Hoare that he thought it would be advisable for him go to New York to describe it to Cooper in person, as it took too long to communicate by telegram. There had been delays in getting messages through as the Western Union telegraph workers were on strike. Finally, McLure told him that Yenser was not going to move the traveller out until he got direction from Phoenixville. Hoare told McLure that was all right and he'd better go to New York and Phoenixville, but before leaving he wanted him to take the levels at the main pier, examine the posts, and see that everything was in perfect line; Hoare wanted to be sure that McLure had as much information as possible on the general condition of the bridge before he left.[8]

Oscar Laberge had started with Phoenix in late 1905. In 1907 he worked on top of the big traveller, doing the rigging where Hall was working. He had heard the men talk about a bent chord in the anchor arm and he'd thought about going to Yenser. Laberge had worked with Yenser on another job and he trusted him. Yenser did good work. Sometimes he got excited, but he also came up with some pretty good ideas. But instead, Laberge went to see his foreman, Aberholdt. The foremen, or pushers, were paid seven and a half cents more per hour. Aberholdt told Laberge, "Yes, that was bent when they put it in there. Don't you remember

TUESDAY, AUGUST 27TH

all the trouble they had to put it in?" Laberge told him that he hadn't been working on the job in 1905 when the chord went in. Aberholdt told him not to worry and to get back to work.⁹

CHAPTER 13

WEDNESDAY, AUGUST 28TH

ON **WEDNESDAY MORNING KINLOCH WAS** out at the front of the bridge about fifteen minutes after seven when he saw that they were loosening the small traveller to run it ahead. He found the assistant foreman and told him they weren't going to move the traveller, but the foreman said that he had orders from Yenser to move it out. Kinloch went looking for Yenser. He met McLure on the approach span and asked him if he knew that they were moving the traveller out. McLure told him yes, they were moving it out. "How about it?" Kinloch asked.

"I don't know," McLure replied, "Only Ben said he had a dream last night."

"That's kind of funny," Kinloch said.

McLure said that Yenser had told him that he had too many men out. Kinloch guessed that Yenser did have too many men out to assign them all to dismantling the big traveller. He couldn't work them unless he was raising steel, which meant that the small traveller had to be moved out.[1]

Yenser had decided to keep going, without waiting for a reply to his letter or to McLure's report. In spite of everything, he had decided on his own to keep his men working. Kinloch knew it would take most of the day to move the small traveller out and fasten it down. No steel could be added to the structure until that was done.

McLure advised Yenser once again that he thought it was "poor policy" to move the traveller out. He also told him that he and Birks had run some calculations and figured that moving the traveller out would increase the stress on chord A9-L by a very slight amount. McLure calculated the increased stress at seventy pounds per square inch and Birks figured fifty pounds. They estimated that this was about one-half of one percent of the stress the chord was already under at that stage. In his own mind, McLure didn't think it would do any harm to move the traveller, and to prove it, he walked behind it while they were moving it out. He examined the chord afterwards and didn't notice any change, but he didn't measure it. McLure also got Kinloch to help him check all related connections and found no further signs of trouble in any other members, other than the ones already reported on.[2]

At 10:30 a.m., the Quebec Bridge Company's chief engineer, Hoare arrived at the bridge site and met with McLure, Kinloch and Birks in the office. He asked if everything had been checked as he'd requested the night before. McLure and Kinloch told him they had inspected all the chords, McLure the upper ones and Kinloch the lower ones. They told him that none of them showed any departure from the normal except A9-L on the anchor arm and C8-R and C9-R chords on the cantilever arm, and that no change had taken place in any of these. Hoare asked whether any rivets or latticing had been broken on chord A9-L. Since the lattices were attached to each rib, if the distance between the ribs was less, as a result of the ribs bending in towards each other, then the lattices should show some signs of buckling. If the ribs were bent when they were assembled to make up the chord, however, then the lattices would not be strained or buckling. But if the ribs were all bent in the same direction, then the distance between the ribs would not change and the latticework would not be affected. Kinloch said the latticing on A9-L was strained; he had struck it with his hammer, and it sang, but everything else was normal. As a result, very little could be determined from the state of the latticing. The surveyor, Cudworth, had taken the levels of the bridge and they were "in exact accord with the theoretical calculations as to what their position would be when carrying its present load."[3]

They told Hoare the small traveller had been moved out. He asked Birks whether he had calculated what effect moving the traveller would have on chord A9-L. Birks said it was approximately fifty pounds per square inch. Hoare said it was a "bagatelle," nothing compared to the stress already on the chord. He then turned to McLure and told him to hurry off so that he could catch the noon train to New York to see Cooper, to lay the facts before him, have a full discussion, and come to some decision about it. He was then to go to Phoenixville and repeat the same explanations there so that there would be no misunderstanding, which might otherwise happen by using the telephone or telegraph.

Birks tried to convince everyone that the chords were bent before they went into the bridge. He reminded them that in June a splice plate had been added between chords A8-L and A9-L of the anchor arm. The plate had been riveted up at that time and now showed no signs of movement or action, either in the riveting or the plate itself, in spite of the fact that an additional three million pounds of weight had been added to the chords. Birks was certain, therefore, that if the bends had occurred after the splice had been installed, the plate and the rivets would have shown obvious signs of stress. Based on this, Birks was convinced that the bends were already in the chord when the plate was put on.

Yenser showed up as Hoare was leaving the office, and Hoare asked him, "So you decided to move the traveller out?"

"Yes, I have," Yenser replied jokingly. "I had a dream, and I think it was foolishness not to move the traveller out." He turned serious then and said, "I have so many men out on the work that I want to employ them."[4]

That morning, Haley and Cook were standing on top of the big traveller. Birks was standing just under where Haley was, talking to one of the workers. "It's all foolishness those fellows talking that way," Birks told the worker. "If Haley hadn't gone down to look at that, nobody would be a bit alarmed."

Haley looked down at Birks and said, "It's perfectly safe, isn't it, Birks?"

Birks hadn't seen Haley; he looked up and smiled. "Why, certainly it is. You fellows are getting alarmed prematurely. There is nothing to cause any alarm."

Haley and Cook told the young engineer they disagreed and went on with their work.[5]

Fifteen minutes later, Yenser was standing with Birks on the top chord, twenty feet away from the big traveller. By then Yenser had ordered the men to move the small traveller forward to the next span. Haley and Cook overheard Yenser say to Birks, "Why in hell don't they let me take down this traveller and get that goddamn load off there before they put up more steel on the end of it?"

The wind was blowing hard, and Haley and the others couldn't hear Birks' reply. Haley thought he knew who "they" were.[6]

Before McLure left for New York, he and Cudworth took some measurements on the bridge, including the masonry of the main pier. Cudworth had installed two benchmarks on each side of the pier. Holes had been drilled in the stone and an iron bar was embedded in lead. They shot the levels with Cudworth manning the instrument, and they determined that the pier hadn't moved—there was no settlement. They then checked the length of the bridge with the use of a 500-foot metal tape, the same one used when laying out the span. The tape was supported every twenty-five feet to reduce sag, crimped at one end, and pulled at the other to correspond to a specified number of pounds at which the tape was standard. The measurements confirmed there had been no movement.[7]

McLure left the site for New York at 12:30 p.m.; he caught the train on the Grand Trunk at one o'clock from the Chaudière Curve. He didn't wire New York or Phoenix to let them know that Yenser had changed his mind and had moved the traveller out. Between Tuesday morning and the time McLure left, no accurate measurements had been made to see if the bends in A9-L were increasing.

Hoare asked Birks to go out on Wednesday evening and examine A9-L again to see whether the deflections showed up at the south splice and extended into the splice plates. The sketch McLure had given him showed the deflections commencing at the splice plates. Birks told him the webs showed slight distortions at the splice between chords 8 and 9. Hoare asked him if he was sure that the lattice didn't show signs of buckling. Birks told him, "No, not the slightest." Hoare remarked that it was rather strange it should be so. The sketch McLure had given him suggested the latticework should show signs of distress, such as buckling. Hoare sent Kinloch to the storage yard to see Clark and get him to refresh his memory about the repairs that had been made to the chord in the storage yard during the summer of 1905. He recalled that the chord had fallen from the grips and that there was a small plate broken and a pair of angles. They had all been repaired from a sketch sent by Phoenixville, which had been submitted to Cooper for approval, at Hoare's request. At four o'clock that afternoon, Hoare left for Quebec City.

Later that evening, Hoare wrote to Cooper, "I have been at the bridge all day trying to get some evidence in connection with the bending of the ribs in this chord. Mr. Kinloch noticed it for the first time yesterday and all inspectors declare that no such pronounced distortion existed a few weeks ago. Mr. McLure made measurements yesterday afternoon and brought them to my house late last night and stated that the erection foreman hastily concluded that he would not continue erecting today, which alarmed me at the time. Upon arriving at the work this morning, he thought better of it and decided to go ahead, at the time asking me if it would be all right. After ascertaining that the effects from moving the traveller ahead and proceeding with the next panel would be so insignificant I requested him to continue, as the moral effect of holding up the work would be very bad on all concerned and might also stop the work for this season on account of losing the men. From further investigation during the day I cannot help concluding that the metal received some injury before it was erected, as the corresponding chord in the same panel, and stressed the same, is in good condition. These panels are being stressed today, approximately, about seven-tenths of their maximum, and it is difficult to believe that this is the entire cause of the distortion."[8]

Hoare's letter wouldn't arrive in New York until Friday. Hoare hadn't gone down to see the chord for himself. He had walked over and looked down at it from the bridge deck above, some fifty feet away.

Haley worked on the small traveller that was used to add sections to the bridge, and so four times a day he walked the entire span. At a quarter past six that night, after quitting time, his partner George Cook took him down to see the bent chord on the cantilever arm that the men had been talking about. With them were Tom Callahan and Harry Briggs. When they got there, they could plainly see that the two outside webs, or ribs of chord C8-R were bulging out from the centerline of the chord and the two center webs were bent into a very long S shape. Haley could see that there had been a big change since he'd last seen it on August 8th when he had walked all the way down the chord to the shoe. He was no engineer, but he could see that there was too much compression put on the chord, that it couldn't stand the strain, and it was giving. He saw, too, that the top lacing that tied the webs together was bent and that the webs of chord C8-R didn't line up with the webs on chord C9-R.[9]

Haley stood on top of the bottom chord and went down on his knees to look over the edge of the chord. He stared at the horizontal line of rivets that ran along the edge of the outside web near the top. He could see the bulge in the web as plain as day. Cook went down far enough along the row of rivets and found one that was sticking out the farthest; half of the head of the rivet stuck out farther than the others. Cook drew a mark around this rivet. The four of them figured on coming around the next day to see if the rivets had been shoved out any farther. Haley told them, "If it's any worse tomorrow, I'm gone from here."[10]

Haley then went over to the other side of the bridge to see if anything was happening with the cantilever chords there. They could see that the outside webs of this chord weren't so much bulging as they were wavy, like a snake, two or three deviations from the straight line that they should have been in. The latticing wasn't broken but some were warped. The splice between 8 and 9 was not yet fully bolted up. Instead of the normal 7/8-inch bolts, in many cases they had to use the smaller 5/8-inch bolts because the holes didn't line up.

An hour later, Haley rode in a carriage to the boarding house with three foremen, Aderholt, Worley, and Slim Meredith. He told them what he'd seen. They all laughed at him.[11]

No iron had been added that day. Dismantling the big traveller and moving the small one out meant that there was actually more steel taken off than was put on.

Birks called Hoare later that night. He emphasized what he had already said, that the chord was bent from the splice, under the cover plate. He had drawn up a sketch and stated that he was positive that the bends occurred in the splice and that some of the bends were there before the chord was put into the bridge. Hoare replied that they would just have to await the result of the McLure trip, and possibly get the answer tomorrow.

Birks mailed a letter to Phoenixville that night, with his sketch. He wrote, "I have made a further investigation of chord A9-L and beg to report following additional data. The bend in the chord starts at the faced splice at the shore end and not at the edge of the splice batten. It appears from this that at least a large portion of the bend was in the chord when the top and bottom splice battens were riveted early in June. This and the fact that the lacing angles are not disturbed leads me to believe that the ribs were bent before erection, in spite of the fact that Mr. Clark and Kinloch think that all ribs were straight when the chord was repaired. From the evidence so far, I do not think we are justified in assuming it to be a fact that the ribs of any of the chords have buckled since erection, and Mr. Yenser has come to the same conclusion."[12]

The letter would arrive in Phoenixville on Friday morning, thirty-six hours after it was posted.

On Wednesday night, Delphis Lajeunesse waited for Haley so he could give him his books for the union. He had been waiting about half an hour and someone said to him, "Just look at the chord." He saw Haley and Cook go down there and after they'd been there a half hour, he saw them going away. He had gone down there that morning, and he thought the bulges were growing, that the chord was getting more crooked. Somebody told him what Foreman Worley had said: "Oh, never mind, we put it in like that."[13]

Charles Davis had heard them talk every day; many of the men thought the chord was buckling. He was driving rivets that day, and on the splice lower down, next to where he was working on the cantilever arm, the jacks were in position to straighten the webs that were buckling. Yenser and Meredith, the riveter foreman, had been down looking at it. When they went away Davis wondered what was wrong. He saw the jacks in between the webs, five to seven joints away from the pier, on the Quebec side. They were trying to straighten the webs so they could get the plate on right. Davis felt uneasy.[14]

No one had done any calculations to determine the actual load chord A9-L could carry in its deflected state, nor did anyone make accurate measurements of the bends in the chord after they were discovered. The chord was allowed to stand for forty-eight hours without any measurement being made to see whether the deflections were increasing. Hoare didn't consider that there was anything dangerous, even though he was personally aware that the chord had gone in there straight. It also never occurred to him to order an inspection of A9-R. He sent Cooper a telegram that day, which read, "Have sent McLure to see you early tomorrow morning to explain letter mailed yesterday about anchor arm chord." He also sent it to Deans.[15]

CHAPTER 14

THURSDAY, AUGUST 29TH

J**OHN SPLICER WOKE UP FEELING** uneasy; the talk from the night before had made him nervous. There were seven Kahnawake workers staying at his mother's boarding house, and they'd told him that there was a place in one of the lower chords where the ribs were bent near the splice. They had tried to jack the ribs together, but they weren't able to, so they'd riveted it up the way it was. He decided that he would lay up for the morning and perhaps go to work at noon; and besides, the wind was up again. It was a clear day, about sixty-five degrees. In fact, the wind was blowing at twenty-three miles per hour—nothing unusual.[1]

McLure arrived in New York at 7:30 a.m. He went straight to Cooper's office and saw Berger, Cooper's assistant. Berger told him the letter he'd sent to Cooper on Tuesday had arrived. McLure sat and waited for the old man.[2]

Kinloch was inspecting the chords early that morning, and he met Birks doing the same thing. They were no longer doing this together as they had done before. Things were now tense between the seasoned inspector and the young engineer. Birks went over to Kinloch and told him, "I think I've discovered where we've made big fools of ourselves, or at least, I think I have anyway. I see what's the matter now. That bend runs up to the field splice." He asked Kinloch if he knew when the splice had been riveted.

Kinloch told him he could take a look at the records and be able to determine the date pretty closely, at least within a day or two. Kinloch then went down to look at chord A9-L again. He found that the chord did show a bend running up, but not as much to the field splice. The number 8 chord seemed to be straight, but number 9 curved into it in some way. The bend that Birks was talking about was near the end closest to the splice, but it wasn't as much as it was in the body of the chord. He didn't measure it and he didn't know whether Birks had measured it or not, but it didn't appear to him that anything had changed from the previous day—the bends were the same, and clearly noticeable.[3]

At 9:00 a.m. erector Theodore Lachapelle was busy taking down the big traveller. The wind was up but nothing out of the ordinary. He just didn't feel like working, though, so he left the job to spend the day in Quebec City.

That morning, Hoare was busy preparing information for the annual meeting of the board of directors, scheduled for the following week. In the afternoon he received a telegram from Deans, "McLure has not reported here; the chords are in exact condition they left Phoenixville in and now have much less than maximum load." After reading the telegram Hoare felt quite comfortable about things. He understood the telegram to be referring to chord A9-L, but in fact, Deans was referring to chords C7-L and C8-L in the west cantilever arm, not to A9-L.

There was a total of 198 men on the project site that day, working on the north and south spans as well as the storage yards on each shore. Eighty-six men were working on the southern span, which projected more than 730 feet from the main pier out over the river. There were seven riveting gangs out; three on the anchor arm, and four out on the cantilever arm. This was one gang less than the previous day as that four-gang had been sent to the front of the structure to help raise the suspended span.[4]

The letters from Yenser and Birks arrived at the Phoenixville office at 9:20 a.m. Deans read the letters and called Milliken, Szlapka, Superintendent Reeves, and Edwards into his office. Yenser wanted to know whether he should continue with the erection. Birks's letter stated that the bend was already in the chord when it went into the bridge. After discussing Birks's letter, Szlapka, the chief designer spent three-quarters of an hour determining what the actual load was on the chord. He concluded that the chord couldn't possibly be bending from excessive stress, since it was only carrying 17,910 pounds per square inch, or three-quarters of the 24,000 pounds per square inch workload for which it had been designed and approved.[5]

THURSDAY, AUGUST 29TH

Deans questioned Edwards on the state of the chord when it left the shop, and Edwards told him that in a number of instances the chords had waves in their webs, but the exact amounts he did not have in his notebook. Deans also discussed it with Milliken, the general superintendent, and he had essentially the same facts. The group came to the conclusion that, "While it was a matter that would ultimately need to be straightened up, the same as other matters, it was not a matter of immediate serious note, and knowing at that time we were going to have a conference with Cooper and McLure, we waited for our final action until after that." They agreed that they would call up the site and tell them to keep going. Milliken got Yenser on the phone and asked him if he had stopped the erection. Yenser said he was going ahead and that everything was all right. He told Milliken that he had moved the small traveller forward the previous morning and that he already had one of the temporary track girder spans in place and was about to swing in the second one. Satisfied, Milliken passed the phone over to Deans. Yenser put Birks on the line.[6]

Deans asked the young engineer if he'd made any further examinations of the chords. Birks told him they'd been watching chord A9-L all of the previous day and that there had been no further movement. He also said he'd examined the chord's lattice angles that tied the ribs together and they showed no signs of yielding; the rivets were tight. He told Deans that after further examination, they were satisfied that either the entire bend or a considerable portion of it had been in the chord at the time of erection. Birks told Deans that the bend extended beyond, or under the splice plate, which had been riveted to the webs the previous June, and that since the plate itself showed no signs of movement or action, this meant that the bends in the chord were there when the splice was riveted. Birks pointed out that three million pounds of weight had been added to the chord since June and it was performing its function in spite of the bend. It was, in Birks view, entirely safe to continue with the erection. He told Deans they had moved the traveller forward and had gone on with the erection. Asked whether he had reported this to Hoare, Birks said he had, and that Hoare had been there the previous day when the examination was made.

Birks made no mention of his discussions with Kinloch, McLure, and Clark, who were certain, based on their own observations, that the chord had been straight when it went into the bridge. As far as Deans was concerned, he had confidence in Birks and felt that Birks was warranted in believing that there had been some bend in the chord at the time it was riveted in June. He told the young engineer to keep an eye on the chord to see how it behaved and that they were going to receive a visit from McLure who'd been to see Cooper in New York, after which they would decide what to do. Deans added, "The people in Phoenixville think you acted wisely in not stopping the erection." The call ended at 10:30 a.m. Deans left the office and caught the 11:09 a.m. train to Philadelphia. He didn't get back to the office until three o'clock that afternoon.[7]

At 10:30 a.m., while McLure sat waiting in Cooper's offices, he received a telegram from Birks. It read, "I do not think we can state positively that chord has buckled since erection; the only definite evidence we have shows the contrary. See my letter with additional data in Phoenixville tomorrow morning."[8]

At eleven o'clock, Birks saw Kinloch on the anchor arm at chord A9-L. "Well it's all right," Birks said. "I've heard from Phoenixville and they have a record that these chords were bent before."

Kinloch laughed at him and told him the chord was not bent when it went in. He turned away and continued his inspection. There was just no doubt in Kinloch's mind that the chord had been straight when it was installed; he had inspected it himself when it left the yard. He simply would not have allowed it to leave the yard with such a big bend in it. Besides, once the chord was installed, it wasn't covered up by any steel; it had been in full view for about five or six months. Anyone walking over it would have seen the bend and reported it. Birks just couldn't or wouldn't believe that the bridge was showing signs of being overstressed. Kinloch went over to the other side and inspected A9-R, and it seemed to him to be all right. Afterwards, he walked over to the cantilever arm, and the chords there didn't seem to be any worse than the last time he'd measured them.[9]

When Cooper reached his office at 11:15 a.m., he exchanged a few words with McLure and went into his inner office to read his morning mail. He read McLure's letter and within minutes he called the young engineer into his office. The old engineer cross-examined McLure—he needed to find out whether "the facts given were actual or whether McLure was scared." Once he realized that the data came from actual measurements, he looked at McLure and said, "It is very serious."

Now McLure was scared. He showed Cooper the telegram from Birks, but Cooper paid little attention to it. In his soft voice, McLure told Cooper, "Mr.

Cooper, they've moved out the small traveller, but we've estimated that it will not add to the strain on chord 9 more than fifty pounds per square inch." He told Cooper that it had been his understanding when he left the bridge that there would be no more metal erected until advice was received regarding the chords.

"Is Milliken on the work?" Cooper asked, thinking he would telegraph orders to Milliken to stop the work.

"No," McLure replied. "Mr. Milliken is not present on the work; there's only a foreman present."

Cooper wasn't sure whether a foreman would take a suggestion from him or not. "I will have to telegraph immediately to the Phoenix Bridge Company for them to wire the bridge. Are you sure that the Phoenix Bridge Company have the same facts before them that you have presented to me?"

"Exactly the same report has gone to Phoenixville that you have now received."[10]

Just then Cooper's assistant handed him a telegram from Hoare; it said Birks had received a telegram from Phoenixville stating that this chord had been bent before it left the shop and that it even had the mark of the chain on it. However, McLure's report stated that the bend in one of the ribs was two and a quarter inches. Could a chord with ribs bent to this degree have been allowed to go into the structure unnoticed? Cooper thought not, but he needed more time to look into it. He picked up his pen and wrote a message to be sent to Phoenixville: "Add no more load to bridge till after due consideration of facts."[11]

He turned to McLure. "Go to Phoenixville immediately and tell them I don't want any delays like we've had on similar occasions in the past. I want immediate action to strengthen that chord and to protect the bridge."

McLure pulled out his timetable—his hands were shaking. He said, "Mr. Cooper, I can't reach Phoenixville before five o'clock."

Cooper added to his dispatch: "McLure will be over at five (5) o'clock." Cooper looked at McLure. "Go!"

McLure ran to catch the train to Phoenixville. It was noon. Cooper's assistant took Cooper's note to the Western Union telegraph office and it was sent at 12:16 p.m.[12]

Cooper's telegram was received in the Phoenixville office at 1:15 p.m. Deans's private secretary read it, attached little or no importance to it, and put it on Deans's desk for him to deal with when he got back from Philadelphia. Deans arrived back at his office at three o'clock. He read Cooper's telegram and arranged for Szlapka and Milliken to be on hand to meet with him and McLure at five o'clock.

Deans ignored the instruction to add no more load to the bridge. Erection of the suspended span continued at the bridge site.

Ingwall Hall had been working on the big traveller for about a month. He would walk out in the morning, return after lunch at 12:45 p.m. and return to shore at 6:00 p.m. He went over the bridge four times a day. He had heard some talk about the bottom chord A9-L on the Montreal side. There had been quite a few men the night before that had gone down there. They had looked up at the deck where Hall was standing and told him that the chord was kinking in from the heft of the wind. He'd also heard his roommate, Harry Briggs, talking about it. That afternoon, the bridge seemed awful springy. They had used the boom on the big traveller to lower sheave boxes filled with iron onto flatcars on the bridge deck. Every time a load was dropped onto the cars, it seemed as though the bridge would spring down about a foot. "It would jar you enough so you would notice it good and plain and you would feel afraid. You would feel the shock every time they dropped anything."[13]

There were now two riveting gangs on the anchor arm; one was working on the A9-L and A10-L joint, and another at the A5-L and A6-L joint. A third was on the main post and the other five gangs were out on the cantilever arm.[14]

Beauvais and his four-gang were still busy riveting the splice plates between chords A9-L and A10-L on the anchor arm on the Montreal side. During the week they had put the scaffold in place, removed the bottom plate, set it down on the scaffold, and installed the temporary angles to keep the ribs spaced apart. They could clearly see the bends on the inside ribs, just like those in the anchor arm on the Quebec side and the chords of the cantilever arm. Beauvais had seen Birks and Yenser examining the chord a day or two before; they'd been there for an hour or more. The splice plates on the outside ribs of A9-L and A10-L were bolted up good, mostly with one-inch bolts, but the plates joining the inside ribs of the chords had hardly been bolted up at all. Out of 280 holes, maybe twenty-five were filled and those with only 5/8 inch bolts instead of the required 7/8 or one inch bolts. Normally, the splice plates on one of the chords would have been riveted up in the shop but Beauvais knew that this was the dropped chord, A9-L. The splice plates on the inside ribs had been damaged by the fall and had to be replaced, which meant that the splice plates had to be field-riveted up on both sides of the joint. The top horizontal plate for the joint hadn't been bolted up either.

THURSDAY, AUGUST 29TH

The inspector, Kinloch, had come down that morning and had seen that the top plate was misaligned at least three-quarters of an inch. They had to use drift pins to line up every hole to get the bolts in. Beauvais didn't like the way the erectors had left things—he didn't like it one bit. He went to see the boss.

Yenser listened to Beauvais's complaint and told him to send a boy to bring down some 7/8-inch bolts and to get it bolted up. Beauvais headed back to the chord and told one of his men to get one of the bolt boys to fetch a box of 7/8-inch bolts. By late morning, they had managed to put in about three-quarters of the bolts. Then they started to drive the rivets, but the inside ribs were bent, badly.

Late in the afternoon, John Norton, Beauvais' partner, found two rivets broken right near the splice on one of the inside ribs. The two rivets were only five or six inches apart. Beauvais had driven them less than thirty minutes before. Norton pulled one out and said to his partner, "Look here, I found it off a quarter of an inch." Beauvais asked how he'd taken it out and Norton said, "I pulled it out." It was broken almost in the centre. Beauvais tested the second one with a drift pin and it was snapped in two. You could turn one end and the other end would be still; it was impossible to pull it out because it was jammed in there. Two snapped rivets—something wasn't right. Beauvais looked at the joint carefully and saw that an eighth of an inch gap had opened up between the splice plate and one of the inside ribs; the plate and the rib were being pried apart. He looked at the top of the two chords where they joined. The ribs on chord A9-L were a quarter of an inch higher than the ribs on A10-L, and the ribs were also bent near the center of the chord. Beauvais called his foreman, Slim Meredith, over. He showed him the broken rivets, the gap between the inside rib and the plate, and how much the inside ribs were bending in at the center of the chord; at least an inch or an inch and a quarter. Meredith looked down into the joint and told Beauvais it wasn't any worse than the others. He didn't think it was serious. Beauvais also pointed out the poor job that had been done with the extra plate that had been added to repair the damage done two years previously. The holes were very bad, and you could hardly get the bolts out. Meredith glanced at it, mumbled something, and moved on. It was 5:23 p.m.[15]

By late afternoon, there were only two sections of eyebars left to be erected on the suspended span, each one weighing about eight tons. Clark had sent one set from the Chaudière yard up to the front earlier in the day and the engine was about to leave with the second set. He spotted Birks as he was about to board the car behind the engine. Clark asked him if there was any truth to what he'd heard, that there was an inch and five-eighths bend in chord A9-L.

Birks stopped and turned to look at Clark. He said, "In spite of the fact that you and Mr. Kinloch think that chord was entirely straight before it left the yard, it's my belief that the chord was in its present condition or nearly so when it went into the bridge." Clark had been the foreman in the yard since the beginning in 1904. He'd been there when A9-L was dropped, and he'd supervised the repairs. There was no doubt in his mind that the chord was straight when it left the yard in 1905. Clark stood his ground. "You have a right to your own opinion, and I have a right to mine. From what I saw and what others saw, I'll retain my opinion."

Birks turned away and climbed onto the car. Clark watched him as he headed towards the bridge. It was getting close to quitting time.

Yenser's men had put in the temporary track stringers first thing that morning and had them in place by eleven. Right after lunch, two sections of the bottom chord of the suspended span were run out onto the stringers. Just before the end of the shift, they had hoisted one chord in straight with the small traveller and it was in position to be bolted up. They had put a few bolts into the splice plates on each side when the first whistle blew.

McLure arrived in Phoenixville just after 5:00 p.m.; Deans was waiting in his office with Szlapka and Milliken. Deans told McLure he'd received a telephone message from Birks. It was Birks's opinion that the bends in chord A9-L were not of recent occurrence and he was writing a letter explaining it all, which would arrive Friday morning. Deans also told the young engineer that he, too, was of the opinion that the bends had been in the chord, maybe not when it was turned out of the shop, but that they had been there for some time. Deans had no evidence to back this up. He told McLure that if he measured some of the chords that were in the Belair yard on the other side of the river, he'd find similar bends. McLure disagreed with Deans. According to Kinloch and Clark the chord had been straight when it went into the bridge, and even Yenser thought the bends were recent.

Deans confirmed that he had received the telegram from Cooper, but he said nothing about Cooper's instructions or whether the site had been ordered to stop

THURSDAY, AUGUST 29TH

the erection. The four men decided to wait until they received Birks's letter the following morning before taking any action.

At that precise moment, chord A9-L was entering the final stages of failure, no longer capable of supporting three-quarters of the load it was designed to carry.

CHAPTER 15

THE COLLAPSE

A T 5:30 P.M. JOSEPH HUOT, the timekeeper, blew the first whistle to signal the end of the shift. The men could start putting away their tools for the day. The second whistle would sound at 5:45 p.m., when the men could start heading for shore. The timekeeper left his office and headed out onto the bridge to check on his workers, as he always did. He'd made a count that morning and there were 117 men working on the south side of the river and eighty-six on the southern span. There were two riveting crews and one hoisting crew on the anchor arm, four riveting crews working on the cantilever arm, and a hoisting crew and crane operators out on the end of the suspended span. The steam locomotive had just come onto the bridge with another eight-ton load of steel. It had just reached the second panel of the cantilever arm.

At 5:31 p.m., Huot was walking on the second panel of the anchor arm, about seventy-five feet out from the anchor pier, when he heard two loud cracks. He looked down and saw that the three-inch compressor pipe that ran under his feet had pulled apart and was running towards the river. The air tank was full and the section of pipe behind him suddenly whipped up sideways—Huot had just enough time to jump out of the way. The second crack was the bridge collapsing. He looked up, saw the portal bending towards the river and knew he was in danger and had to escape. He turned around, jumped, and ran up the hill to make the

THE COLLAPSE

approach span. As his feet were striking the planks on the bridge deck, they were sinking under him, disappearing.[1]

Alexander Beauvais was standing on the scaffold underneath the splice between chords A9-L and A10-L of the anchor arm on the Montreal side. He was up inside chord A10-L, between the inside and outside ribs, a space of less than two feet. He was riveting the splice plate on the inside ribs. He was about to drive another rivet when he felt the chord drop. He heard nothing. "As soon as I felt it break, I made a grab for the splice plate. I had my arm on the plate. I just turned my hand out and caught the plate. There was a space of an inch and a half and I got my hand in it." He let go of the heavy riveting gun he'd been holding. It fell and shattered the bones in his foot, but he felt no pain. When the chord stopped moving, it was suspended three or four feet in the air, with him still inside it. He held on to the chord with one hand and never touched the ground. When everything was still, he came out. His nose was broken but being inside that chord had saved his life. Norton, his partner, didn't make it, neither did the rest of his four-gang.[2]

Delphis Lajeunesse and his brother were bolting on the main post of the anchor arm, on the Montreal side. The twenty-four-year-old worker was standing on a sheave box putting a turn on a rope to send a box of bolts up to his brother. The whistle had just blown and his brother told him he had no time to send it as it was pretty near quitting time. Delphis felt something jerk the bridge. "I fell down on my box, stood up, fell down again inside the box, and I looked again." Suddenly, he was thrown six feet onto the box brace that ran diagonally between the chords on either side of the bridge. "I thought the traveller had fallen down on the bridge. The traveller was in the same place. I came to this side of the bridge and I looked, and when I saw the bridge go down in that way, I was on that chord, and I thought that chord made the bridge fall." He was standing on chord A9-L.

Delphis turned around and saw that the brace was moving towards the Quebec side. He thought, *Well, I'm finished.* He watched as the post came down and he thought it was coming down on top of him, but it didn't. Nothing landed on him. He was the first to come out, and then his brother. Delphis hadn't been thrown from the bridge; it had caught him, saving his life, but his left leg was broken.[3]

Delphis's brother Eugene was on the floor beam of the anchor arm, above where his brother was. He had just told Delphis it was quitting time. He felt a jerk and was thrown upwards; he came down hard across a stringer. He said to himself, "I'm finished." He went down with the stringer as it hit the ground hard, but not too hard. He blacked out for a moment. When he came to, he got up and walked away from the wreckage. He looked over his shoulder and saw his brother—he, too, had been caught by the bridge.[4]

James Johnson was in charge of the bull gang, the labourers handling the iron. He was standing with Joe Lefebvre on the ground about fifteen feet west of the anchor arm when he heard the locomotive run out on it. He heard a racket like a piece of iron falling off a car, or a car jumping the track. He looked at the third panel from the pier on the anchor arm and saw what he thought was a stringer or a chord parting. He then looked up towards the main pier, saw it coming, and he ran. He looked back and saw the whole thing wave, or kind of rock. He ran thirty or forty feet and looked again and the only thing he could see was a cloud of smoke. Above the smoke he could see the wooden false work on the pier fall back towards the shore. Everyone else in his bull gang was busy running out of the way except for one, a boy. He'd been working on some pins directly under the anchor arm, earning ten cents an hour. He was crushed. The boy was fourteen.[5]

Ingwall Hall was working at the top of the big traveller that was on the tenth and last panel of the cantilever arm, four hundred feet above the water. One of the booms on the Montreal side of the big traveller was hoisting up two large timbers on top of the girders. Hall was on the Quebec side, facing south towards the main pier. He couldn't see much as the main steel girders at the top were three feet from his face. Suddenly, "I could feel it start to go down; it seemed just like it was tipping on an axle, like on the pier. It was going down so fast you got tears in your eyes, and you could hardly realize anything beside you. My partner was just about seven or eight feet from me, and I never noticed him and never saw him—never knew anything. At first it did not make an awful noise when it started. It went fast at the start till the deck of the bridge struck the water, then it seemed that it kind of slowed up a little bit. I went under and when I came to the surface, everything was out of sight except timbers, and I don't know how many voices were hollering for help. The water was too unruly for me to notice how many. It seemed it was going fore and back in small waves so you would have to hoist yourself up to the chest to breathe without drinking water."

Hall had remained in his exact position on the traveller as it went straight down—he never lost consciousness. He'd lost two fingers and the flesh off a third, but he was able to swim to shore.[6]

Hall's partner, Oscar Laberge, was standing on top of the big traveller with his feet planted on two pieces of timber, ten by twelves, thirty-eight feet long. Suddenly, "it started to go down. I just touched the timber; my feet just lightly touched the top of the timber. The balance, the iron, was going faster than the timber, and the timber was going about the same as I was, and I was just touching them so I could not lie down to catch it because I was going too fast, and I was standing up, and when we were down seventy-five or a hundred feet, I kind of stopped, so I remember my feet caught the timber, and as I was going down I had my arm around one of those pieces of timber. I remember I was in the water. When I hit the water, I do not know where I hit or anything like that." As soon as he came to the surface, he swam as hard as he could and caught up to another timber. He held on to it, and after a few minutes, a skiff came by. The fellow in the skiff lifted his leg in and took him to shore. Laberge's jaw was broken and his pelvis was shattered.[7]

Percy Wilson and another labourer, who'd been on the job for just three days, were distributing cold rivets to the seven riveting gangs on the structure. Wilson was walking on the approach span going ashore for more rivets. He heard a noise, turned around, and saw it going from the far pier like a flash of lightning. "It took about five or six seconds and then all I seen was floating timber on the river and a mass of steel was between the two piers." He found himself running towards shore with Ouimet and Huot. He ran down the stairs next to the shore. One of Wilson's brothers was working on the traveller and his second brother was working under the anchor arm. "I thought I would see him down below, and when I got down there, I did not see him— they were gone, both of them." The labourer working with Wilson that day was also killed.[8]

Charles Davis was as far out on the bridge structure itself as you could get. He was on the suspended span, standing on the bottom chord section they had just lifted into place; the erectors had a few bolts in on each side. Suddenly, "I heard a crash, something go away back on the bridge, and I felt it sink." He just went straight down. "It left me. I was in space, in the air. It travelled a great deal quicker than I did." He looked down and a good many thoughts went through his mind. "I don't remember anything striking me at all." He hit the water and blacked out. Someone found him and took him to shore. But he had been struck by something on the way down; he suffered serious injuries to his back and hip.[9]

D.B. Haley, the president of the union and signalman on the small traveller, was working on the top chord of the small traveller, where he'd always been since he'd started in June. The traveller was sitting on the third panel of the suspended span. Haley was on the end of the jib that extended out from the small traveller. He was out over the fourth panel of the suspended span, past the structure as far out over the water as anyone could be. "I was on the extreme end of it and the first thing I knew, I caught myself going through the air. What I was sitting on fell away from me and I fell through space. I realized that the iron fell very much faster than I did and left me going through the air. The next thing I remember I was deep in water." He came up, climbed onto some planks, and was soon rescued by a boat that had come from the other side of the river. He didn't see the structure falling behind him, and he didn't hear a thing. He was in the water before the noise came.[10]

J.J. Nance was running one of the electric hoisting engines at the top of the small traveller. Haley was on the jib giving hoisting signals to the two men on the engines. "She went down so quick, you didn't have time to think of but very little. I went down from the top. I knew very little after she started, and the next second, she was in the water and I was down in the water with it. I went to the bottom with the engine." Nance was struck by the wreck, suffering two broken ribs and a pulled muscle.[11]

Michael Esmond was a boatman, also known as a lifesaver. If anyone fell into the river, he was there to pick them up. He had never been on the structure. He had run underneath the bridge two or three times that day. He ran a line from the main pier and attached a small buoy at the other end, fastening the buoy inside his boat for quick release. He was floating on the ebb tide, a hundred feet downstream of the bridge, just sitting there waiting and watching. "My attention was all the time on the rising gang, the climbers, the men working outside." He was even with the big traveller, about 600 feet from shore. The small traveller was on the outside, putting on the iron. Esmond was looking up at the men when he heard a heavy noise at the shore, "like a clap of thunder." He looked and "I saw everything going. It seemed that it all went in a body." He touched the buoy and watched as the big traveller fell towards the river. "It seemed to me to be going out, because if it had come east it would have taken me for sure." When it struck the water, "It made an awful swell. It came right straight to me and I didn't know where I was. I sat right down there, and the boat headed for the east. The next wave was not as large and when it cleared up, I heard men shouting for assistance and I went to get them out." He didn't know how he had cleared the buoy, but he had.[12]

C.L. Culbert hadn't worked that day. He was out walking along the south shore with another worker, Richard Chase, who had also taken the day off. The two men were discussing how many more panels were left to be added to the suspended span. Culbert told Chase, "There ain't any more." Suddenly, Culbert saw a light at the very top of the anchor arm, like a flash of electricity. Without taking his eyes off the structure, he said to Chase, "There she goes." They started running for the bridge. The anchor arm seemed to rise up a little and then it just began to crash and rumble. He saw someone up on the bridge running towards the shore. He looked to his right and watched the cantilever span hit the water. He got there as fast as he could to see if he could get anybody out who was crippled or injured. He saw a boat sitting at one side and his idea was to get that boat and get it into the river.[13]

As he was walking with Culbert, Chase was looking down and had his hands in his pockets. He heard a noise and looked up at the towers and the span out over the river. He saw the tops of the towers slowly lean out over the river twenty or thirty feet, the lower parts suddenly kicked backwards off the main pier, and it all fell in a heap. He started running with Culbert and had to watch for the stones at his feet. He was looking down as the cantilever hit the water, but he looked up in time to see the big traveller topple over into the river. He watched as the cantilever sank. It seemed to have kept its whole length when it went into the water. He saw the locomotive. It was moving and was one panel out on the cantilever arm, fifty feet from the pier. It shot one more panel out towards the water and went down with the bridge.[14]

The engine-runner on the steam locomotive was thrown into the river and was rescued, but the other men went down with the bridge. Birks, the young Phoenix engineer who had been riding on one of the cars behind the engine, went down with the bridge and perished.

Frank Cudworth, the resident engineer for instrument work, was in the office on the opposite side of the river, about half a mile away. His attention was first attracted by an unusual noise. He thought at the time that a steel plate had dropped. As he was turning to look out the door, the noise continued, and he knew there was something wrong. His eyes were drawn to the top of the main post and the post peaks. The posts were falling, at first, slightly towards Quebec, then more so towards the river. Then suddenly, these two motions stopped, and the posts went straight down together, as one unit. "They just seemed to sink out of sight." It all happened in seconds.[15]

Kinloch was at the Phoenix office, which looked straight down the centerline of the bridge. His attention was called to "a noise, like a car running over a stick of timber—the crunching sound of timber—not very loud. In fact, I wouldn't have paid any attention to it if it hadn't continued. I was just entering the office of the Phoenix Bridge Company and, with the noise continuing, I looked out from the door and saw the end post on the Quebec side trembling. I knew something was wrong and I stooped down a bit and looked up through the glass in the door to see the portal. It was inclined slightly away from me and trembling. The posts were leaning over and slowly sinking. I stepped outside, and the same motion was continued. It was slowly sinking; the two center post peaks were slowly settling straight down in about the same position that they always stood in regard to the line of the track, just the same as if they had been ice and were melting off at the bottom. They didn't seem to be east or west towards Montreal or Quebec, they seemed to be about the same as the rest of the bridge. About that time, I turned my back to it and didn't look at it anymore." The portal had inclined towards the river. When he looked again, it was all down. He turned his attention to the approach span and looked to see if anything had happened to the legs of the span. He knew there would be a lot of people there, and "I didn't want it to go down on top of them." When the dust settled, he could see there was nothing wrong with the approach span, but the anchor arm lay in a twisted heap between the piers.[16]

Ulrich Barthe, the secretary of the Quebec Bridge Company, had been touring the project with some visitors, and they were on the south shore when the bridge fell. He recalled later, "There followed a great silence, profound and mournful, as if all of nature was hushed and repulsed in horror."[17]

The fall had taken less than fifteen seconds. Thirty million pounds of steel were piled in a twisted heap along the foreshore and in the river, some of it in two hundred feet of water.[18]

There had been many men working on the anchor arm, between the piers on the shore of the river. The bridge collapsed at low tide. The tide began to rise at six o'clock. Rescuers were able to pull eight injured workers from the wreckage, and they were rushed to the nearby Lévis hospital. Eleven other survivors could not be pried loose from the massive steel tentacles. Those who came to the rescue tried frantically to save the men, but there were no cutting torches or any heavy tools available; they struggled in vain with crowbars and tackle. One worker was pinned by the foot in the center of the wreckage; he shouted to the rescuers to chop off his foot with an axe and get him out, but they couldn't reach him.

The tide continued its inexorable rise of thirteen feet. From the time the sun set until the moon rose at ten o'clock, the rescuers were aided only by the light of a few lanterns and bonfires that had been lit along the cliff. A crowd had gathered at the top of the steep riverbank 150 feet above, among them relatives and friends of the trapped workers. The horrifying scene below would be with them forever, most of all the cries for help—some of the men pleaded to be put out of their misery. The enormity of the catastrophe silenced those watching. A priest from the nearby parish of Sillery was in the crowd; Father McGuire lowered himself down the cliff on a rope and waded out into the river through the twisted steel wreckage to administer last rites to the doomed men.[19]

PHOTOS

All but the last two photos were entered as exhibits during the Royal Commission of Inquiry.

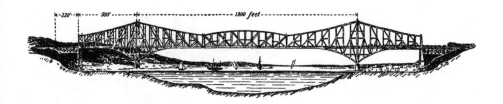

The Phoenix Bridge Company submitted a sketch of its cantilever design to E.A. Hoare in 1897. Originally planned with clear a span of 1,600 feet between the main piers, the span was later changed by Cooper to 1,800 feet, making this the longest clear span bridge of any kind in the world.

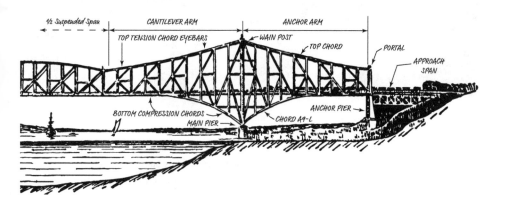

(2) Lower chord A9-L of the anchor arm in Phoenixville, ready for shipment to Quebec in 1905.

A9-L is shown here lying on its side, with a small transverse strut lying on top of it. The six splice plates at the end of the chord are just visible inside the temporary box built for shipping. The chord was destined for the western truss on the Montreal side of the bridge. Note the lattice angles tying the four ribs together. Chord A9-L is linked to chord A9-R in the east truss by means of diagonal transverse struts that are riveted to the large plate that has the writing on it. Chord A9-L would ultimately initiate the collapse. (Library and Archives Canada)

PHOTOS

(6) View of an anchor arm lower chord and splice plates. Date August 11th, 1905. Note the arrangement of the seven splice plates.

This compression chord is resting on the bridge deck prior to placement into the structure. There are six vertical plates at the end of the four ribs- one on each side of the two outside ribs, and a single plate on each of the two inside ribs. (Library and Archives Canada)

(17) View showing progress. Date November 23rd, 1905. Erection for the season is completed.

The lower chords of the anchor arm are in place and are resting on temporary shoring. The large traveller, positioned at the fifth panel of the anchor arm, has nearly completed the erection of the sixth panel of the upper truss. Lower compression chords A9-L and A9-R are in the ninth panel of the bottom truss, one panel to the right of the main pier. (Library and Archives Canada)

PHOTOS

(14) View showing the raising of a set of top chord eyebars. Date June 20th, 1906. Note the various tackles.

Perched nonchalantly atop a bundle of eyebars more than three hundred feet above the river, these workers are about to join the top chord eyebars to the main post with a steel pin. Within one week, the entire south anchor arm will be in place. (Cover photo) (Library and Archives Canada)

View of the lower chords after the shoring has been removed. The riveting of these chords was just getting under way. (Library and Archives Canada)

PHOTOS

(16) View showing the erection of a cantilever arm lower chord. Date July 5th, 1906. Note the latticing of this chord which is No.10 in the right truss.

The chords of the anchor arm were nearly identical to this cantilever arm chord. (Library and Archives Canada)

(28) View showing progress. Date Novr. 29th, 1906. Erection for the season is completed, the Cantilever Arm being practically finished. Note the steel falsework.

The shoring was removed at the start of the 1907 season, at which time the riveting of the lower chords could begin. (Library and Archives Canada)

PHOTOS

(5) View of the upper chords and lateral bracing panels 5 to 10, Cantilever Arm.
Date August 15th, 1907.

The Quebec Bridge Company's inspector, Kinloch, climbed 400 feet to the top of the big traveller to take this photo. This view looks south. Note the two men standing between the ornate finials atop the main posts. (Library and Archives Canada)

(35) View showing progress. Date August 28th, 1907. Note that the little traveller has been moved forward into position for erecting the fourth panel of the Suspended Span. Note also the condition of the big traveller.

This was the southern span the day before the collapse. The view is from the south shore of the river looking downstream towards Quebec City. Note chord A9-L. At this stage, the structure cantilevered 730 feet out over the St. Lawrence. (Library and Archives Canada)

PHOTOS

(24) General view from the anchor pier northwards. Note the unbroken lines of the upper chord.

The bridge collapsed at 5:31 p.m. on August 29th, just minutes before quitting time. It took fifteen seconds for the structure to fall. This photo was taken the day after, at low tide. Cooper's top tension eyebars are draped over the wreckage, with only one broken bar. Nine thousand tons of steel lay between the piers in a twisted heap 80-feet wide, 500-feet long, and up to 40-feet high. It's clear from the wreckage that the structure went straight down. The cantilever arm and suspended span are in the river, two hundred feet below the surface. (Library and Archives Canada)

(21) General view looking towards the main pier. Note the upper chord in the right.

This is what the rescuers had to contend with. "Bending, crushing, buckling, twisting, shearing, and tearing..." is how *Engineering News* described the wreckage. It would take two years, twenty tons of dynamite, and 90,000 cubic feet of oxygen to remove the debris. (Library and Archives Canada)

PHOTOS

(20) General view of the main pier southwards. Note the moderate damage to the upper laterals.

This was taken at low tide. Within minutes of the collapse, a crowd began to gather at the top of the steep riverbank. (Library and Archives Canada)

(12) View from the West showing the beginning of the first 180° bend in chord A-9-L.

The four ribs of the chord ended up in the shape of the letter S, consistent with the bends they exhibited prior to the collapse. (Library and Archives Canada)

PHOTOS

View looking north. The north main pier and shoring are visible on the opposite shore.

The tide is rising. Of the seventy-six men who perished, only thirty-eight bodies were ever recovered. Most men were trapped in the twisted steel of the cantilever arm and suspended span that are still at the bottom of the St. Lawrence. Although the two piers on the south shore survived the collapse, they were not used for the second bridge, and neither were the two piers on the opposite shore. The granite facing stone was salvaged but the piers were demolished to low water level. The new bridge is located sixty-five feet south from the original site, and is twenty-one feet wider. (Library and Archives Canada)

(1) Shows buckling of chord A1R between cover plates; also efficiency of covers and diaphragms in keeping members straight. The chord apparently struck the ground end on and the buckling shown resulted.

This photo captures the failure mechanism of the compression chords. The riveted latticing between the splice plates has failed. Unrestrained by the lattices, the four ribs could no longer act as a unit and, "buckled together like so many sheets of paper crushed in the hand." (Library and Archives Canada)

PHOTOS

On September 11, 1916, nine years after the collapse of the first failed bridge, thirteen men lost their lives in what proved to be an accident during the construction of the second replacement bridge. A metal casting broke, releasing a corner of the center span, leading to failure of the other three. The center span still rests at the bottom of the St. Lawrence, alongside the suspended span and cantilever arm of the first failed bridge. (Library and Archives Canada)

A photo of the bridge as it stands today. The superstructure on the second bridge was over two-thirds heavier (66,480 tons vs. 39,000 tons) and made of nickel steel, which is lighter and stronger. The chords on the second bridge were straight and had a cross-sectional area twice those of the first bridge (1902 vs. 842 sq. in.). The first bridge looked almost flimsy in comparison. With a clear span of 1,800 feet, it reigned as the longest span bridge of any kind in the world until 1929, with the opening of the Ambassador Bridge between Windsor and Detroit. It is still, however, the longest cantilever bridge in the world. (The National Trust for Canada)

CHAPTER 16

THE DEVASTATION

M CLURE HAD LEFT THE PHOENIXVILLE office just after five-thirty. Fifteen minutes after he left, the phone rang in the Phoenix Company's office and Milliken answered. It was Waitneight, their chief timekeeper at the bridge site. The connection was bad, and Milliken couldn't understand what the man was saying, but he knew something was wrong. His hand was shaking as he handed the phone over to Deans and went looking for the Phoenix Company operator. They tried but couldn't get a good connection. Finally, Deans called the manager of the phone company, and they got through at ten minutes past seven. Their worst fears were confirmed; the bridge had collapsed and nearly all of the men working on the southern span had been killed, including Birks and Yenser. Peter Szlapka was in his office; when they told him, he collapsed.[1]

John Montour, a young riveter from Kahnawake, had been sent off the bridge for food for his riveting crew because they planned to work late. Montour heard the crash and felt the ground shake under his feet. He stopped dead in his tracks. It was bad—he knew it. He turned around and ran back to the riverbank. As the dust began to settle, he looked down at the horrible scene below. Almost fifteen thousand tons of structural steel that had once towered hundreds of feet above the St. Lawrence River had disappeared or lay in a mass of tortured metal between the piers. The only parts of the bridge that were recognizable were the long eyebar tension members of the top chords; they lay unbroken, draped over the tangled wreckage and the main pier, and then disappeared into the deep water with the cantilever arm and suspended span.

Montour knew he had to do something, but he didn't know what. He started running towards the wreck—he had to find out how many of his people had been hurt. He tried to get in close to help but they wouldn't let him near it. The young native went to look for another Indian, and he finally spotted Joe Regis from his village. The two of them searched frantically for a telephone to call home. They reached the postmaster at home in Kahnawake at six-thirty. Antoine Giasson had trouble understanding what Montour was saying at first and when he finally understood, the enormity of it hit him and he dropped the phone and ran outside. His anguished cry could be heard throughout the village. "A disaster, a disaster, a horrible disaster!" The villagers ran to Giasson's house; he told them, "The Quebec Bridge has collapsed, all of our men are dead."[2]

The terrible news spread in seconds; not a single family had been spared. Thirty-eight ironworkers from the village had been at the bridge site. Thirty-three of them had perished and two were injured. They left behind twenty-four widows and fifty-two orphans. The hardest hit was the family of the elder Pierre D'Ailleboust. He'd lost seven members of his family; his four sons, his brother and his son, and his brother-in-law. One worker, Orite D'Ailleboust, left behind nine children and his pregnant wife. All seven Indians from Kahnawake who had been staying at the Splicer boarding house and who had talked about the bent chords the night before the collapse, were dead. The others in John Montour's four-gang had perished as well; the bodies of his brother, his uncle, and his cousin would never be recovered.[3]

McLure didn't hear of the collapse until seven thirty. He went looking for a telephone to call Cooper. When McLure told him that the bridge was in the river and that many of the men had been lost, there was dead silence on the other end. McLure wasn't sure if Cooper had heard him say that he would see him at his office in the morning, before returning to Quebec. The next morning, when McLure walked into Cooper's office, the old man exclaimed, "Well, it's that chord!" McLure left a few minutes later. The old man was "not feeling very good."[4]

The morning after the collapse, a group of thirty people left the village of Kahnawake for Quebec. They went to bring their loved ones home. Only eight bodies had been recovered. At 9:15 a.m. on Sunday, September 1st, the train

THE DEVASTATION

carrying the eight caskets arrived at the Adirondack station near Kahnawake. They were unloaded one by one at the station, and the procession made its way slowly to the village. The caskets were taken to the homes of the victims for a wake that would last through the night. The funeral procession on Monday morning stretched half a mile. With only six hearses available, two of the caskets had to be carried by hand. The church grounds were filled with hundreds of people, mostly strangers, there to offer support to the devastated community. As the procession arrived at the church, a strong breeze unfurled the flag flying at half-mast in front of the town hall across the road. The Archbishop of Montreal, Monseigneur Bruchesi, presided over an emotional funeral service, and an Iroquois choir sang liturgical chants in their native language. The eight caskets were then taken to the churchyard cemetery and buried in a common grave. As the mass grave was filled, two Iroquois women sang a chant for the dead.[5]

Many of the workers had come from the two towns closest to the bridge site. A long funeral procession made its way through the streets of New Liverpool and Saint-Romuald, carrying the bodies of three members of the Hardy family and sixteen-year-old Wilfrid Proulx, the only bodies from the two communities yet recovered. Their caskets were taken to the parish cemetery and buried in a common grave.

The body of twenty-six-year-old assistant foreman Chester (Slim) Meredith, the only American victim yet recovered from the wreckage, was sent home to Columbus, Ohio. The funeral for the young ironworker was attended by many members of the local branch of the Bridge and Structural Iron Workers' Union.

The Phoenix Bridge Company offered a fifty dollar reward for every body recovered. The St. Lawrence River teemed with small boats searching for bodies; the search lasted for weeks. John E. Birks, the Montreal jeweler, offered a $300 reward to anyone who could find his nephew's body. On September 11th, two young boaters found the body of Arthur Birks near Cap Blanc, three miles downstream. Three bodies were found as far as L'Ile d 'Orleans, twenty miles from the bridge site. One of these was that of American ironworker Joseph Ward, who had fallen from the bridge a week before the collapse, and for whose body a reward of $200 had been offered. By mid-September, the families of thirty-eight of the seventy-five men who had been killed had some closure; the remaining bodies were never recovered.[6]

The headline in the *New York Times* read, "Bridge Warning Was Just Too Late." Cooper told the reporter from the *Times*, "Thursday morning my inspector came down to my office and told me that things did not look well for the bridge. He thought that it ought to be looked into. Immediately I wired the man in charge of the work there to get off the bridge at once and stay off until it could be examined." In a subsequent interview, Cooper clarified his previous statement. He said that he had no authority to order the men off the bridge. "I did not have that right; I am the consulting engineer only, and upon the report of my inspector, I wired to the offices of the Phoenix Bridge Company, at Phoenixville, Penn., and warned them. I had no right to wire the men actually at work on the bridge."[7]

The *Montreal Daily Star* interviewed William Reeves, the Phoenix general superintendent, about Cooper's telegram. "Owing to the telegraph strike," Reeves told them, "the message from Cooper to add no more weight to the structure wasn't received until the middle of the afternoon on Thursday. Deans, the chief engineer, to whom the message was referred, was absent from his office at the time it was received. It was put on his desk. He came in at five o'clock, and fifteen minutes later, McLure arrived with particular information about the bridge. It was then too late in the day to get a message through to the workers before they quit work." According to Reeves, "The fact that the message was not explicit in advising the calling off of the men on the structure is the reason for the failure of other officials to act at once during the absence of Deans. There was no suspicion of any immediate danger. There was no negligence on the part of anyone in the matter; the message was not considered an urgent one at all. Had there been any thought of the lives of the men being in danger, every effort would have been made to have the work stopped at once. The work has been under the closest supervision all along, and one of our engineers had just returned from there with a favourable report. Since the telegram arrived only a couple of hours before the accident took place, it is not likely we could have gotten word through to them in time anyway, but we would have tried had we dreamed it was necessary."[8]

Reeves was referring to Szlapka's visit to the bridge site the previous week, during which time Szlapka had not examined the bent compression chords. Szlapka was not in a position to give a "favourable report." Reeves told *The Gazette* he couldn't imagine what had caused the collapse of the bridge.

Cooper spoke on the phone to David Reeves, the president of Phoenix on Saturday, two days after the collapse. He told him that he had no idea at the time there was any immediate danger, nor could he account for the actual failure.

The *Montreal Daily Star* also reported that Charles Connard, private secretary

THE DEVASTATION

to Deans, at first denied any knowledge of the telegram, but finally admitted that it had been received. "It came late in the afternoon, and gave no warning to call the men off, so we held it for Mr. Deans. Comparatively little importance was attached to it. Mr. McLure got here almost as soon as it did. I do not know what time it was dated in New York, but I think there had been some delay in transit. No blame can be attached to anyone regarding the telegram."[9]

On September 2, the Quebec Bridge and Railway Company chief engineer, Hoare, wrote to Cooper to "Correct a misstatement" in his letter of August 28th, "which was written late and very hastily." He wished to revise his letter to make it clear that he had not requested that the erection work be continued but that he had, instead, merely acquiesced to it. "As stated in my last letter, strictly speaking, I did not request the foreman to continue the work, as he had already done so. At the same time, we thought there was no immediate danger in adding so small a load."[10]

Ben Yenser's sister had come from Philadelphia to Quebec to visit her brother and sister-in-law. In the early afternoon of the 29th, she had boarded a train in Lévis to return home through Montreal and New York. When she arrived in Montreal at 7:00 p.m. she went to the Pullman office to purchase a berth on the train to New York. When she gave her name to the agent, he told her that he had a telegram for her. It was from her sister-in-law, Sallie, and it read, "Come back at once. Accident at bridge."

Yenser's sister told the agent, "Oh, I have to go back to Quebec; something terrible has happened. Can you give me a berth on the next train to Lévis?" While the agent was making out the reservation, another telegram came, telling her that her brother had perished. Dropping her umbrella and handbag, she swayed against the railing in front of the ticket office. She was helped to the waiting room to wait for her train and left forty-five minutes later, arriving in Lévis the following morning.

On Friday morning, the man who had gone to Boston to hire more men, wired the Phoenix office that he had secured the required men. They wired him back and told him to hold the men and report to Quebec before securing any more.[11]

On the Tuesday following the disaster, Deputy Superintendent General of Indian Affairs Canada, Frank Pedley, Esq., sent a letter of sympathy to the mayor and council of Kahnawake. With the help of an Indian Agent, Chief John Sky replied the next day on behalf of the Band, asking the Department to, "Take in hand the cause of the widows and orphans" to obtain damages for the thirty-six victims, including three men who were injured. Pedley dispatched one of his most experienced inspectors to travel to Kahnawake to meet with the Band. As Inspector of Indian Agencies and Reserves, James Macrae had many years' experience dealing with Aboriginal claims from treaty and non-treaty Indians for land and money scrip, primarily in the Western Territories. But he had never dealt with anything like this.[12]

The Federal Department of Indian Affairs opened Red File 316880 for "The Quebec Bridge Disaster" and its impact on the Kahnawake Indians, and more particularly on the "Claims of Dependents (Lists of Killed and Injured, clippings)". The first clipping, titled "Revised List of Caughnawaga Dead," was from *The Montreal Star*, dated September 13, 1907. The subtitle stated that "Thirty-six Indians were at the bridge, thirty-three have perished." The Star reporter had worked with the Reverend Father Forbes and the present pastor, Father Granger, to put the list together, which identified the dead, eight of whom had been found. The list also included their marital status as well as the number of children they had. The article made special note of Louis Deer, who had left his village just the previous day and was only a few minutes from completing his first shift.[13]

One week after the collapse, Macrae travelled to Kahnawake and sat down with the band's chief, a number of band members, and the village priest, Father Granger. Communicating through a French/English interpreter, he offered the department's assistance in arranging for "concerted action" to assert any claims for insurance, damages, or other benefits. In response to the question of whether the department would assume responsibility for any damages, Macrae told them the sufferers' claims were of a private character, for which they had recourse in law, and that no responsibility could or would be assumed by the department. He explained that the department could only act as an adviser. The inspector recommended that the sufferers do nothing in haste. They were not to enter into any contracts or settlement agreements, without full consideration and reliable advice. Father Granger told Macrae that everyone had decided to wait for the results of the investigations before making up their minds. He said that he and the people on the reserve had great confidence in the generosity of the Phoenix Bridge Company and thought that they would extend "liberal treatment" to all sufferers.[14]

A suggestion by Macrae to engage legal counsel to watch the progress of events was turned down by the band. One councillor asked the inspector whether it would be appropriate for the band to use some of its funds to assist the widows and orphans. Macrae told them they should do so only if there was "real need," since future claims for damages might be affected. Another suggestion by a band member to distribute ten dollars a head to assist the widows and orphans was also dismissed. While he was on the reserve, the inspector had seen the paymaster for the Phoenix Bridge Company; he was there to pay wages due to the families of those who had been killed or injured. The priest told him that fifteen days' wages were due and paid.[15]

Father Granger then stated that the only really pressing matter was to relieve the widows by taking their children and placing them in industrial schools. He knew of a number of vacancies at a school in Manitoulin, Ontario. Wikwemikong was an institutional school for aboriginal children run by the Jesuits and funded by the federal government—the institution was known for its strict discipline. Father Granger promised to send Macrae a list of the children which numbered "about eighty." Macrae said he would give the matter earnest consideration and promised to present the priest's views to his boss, Pedley.[16]

Macrae met with Pedley the following day in Ottawa and told him of his meeting with the band. Pedley and Macrae shared the same reservations about sending the children away to industrial schools. Pedley thought that this might be seen as a mitigation of damages, thus reducing the claims of the victims. The two men decided that the children should remain with their families in the community, at least for the time being.

The band and its clergy persisted, however; minutes of a meeting held the following day, September 6th, showed a resolution by the Band Council to ask the department to place fifty or sixty orphaned children in the Manitoulin school. The department responded with a request for a list of the children that the Band proposed be sent to Wikwemikong. Macrae received the list and realized that only a small number of the children listed had in fact been affected by the tragedy.

Macrae travelled to the reserve on October 23rd and spent two days assessing the condition of the widows and orphans of those killed. Macrae reported that the union, the International Association of Bridge and Structural Workers of Indianapolis, had paid the workers' death benefits of $100, "in the most prompt manner." Wages due and funeral expenses were also promptly paid by the Phoenix Bridge Company. A relief fund collected $8,000 for the families of the Kahnawake workers. In his report to Pedley, Macrae wrote, "The amount of insurance carried by the married Indians is noteworthy. We know of $19,190." He concluded his report: "I sent for one or two of the widows whose cases it seemed might call for immediate personal attention and found that such was not now necessary. If it becomes so, they are to communicate their requirements. There is not present pressing occasion to place any children in industrial or boarding schools solely as a means of relief." The children remained on the reserve with their families.[17]

The day following the collapse, a separate relief fund was set up for the widows and orphans of the dead workers from nearby communities, including St. Romuald and Liverpool, as well as for the parents of the single workers who had been supported by their sons. By September 18th, the fund totaled $11,569, at which time the collection ceased, the funds were distributed, and the relief committee was disbanded.

The first lawsuit for $15,000 was filed against the Phoenix Bridge Company by the father of Zephirin Lafrance for the death of his eighteen-year-old son. As of September 6th, three more lawsuits were launched on behalf of the dead workers. By September 9th, eight lawsuits had been filed seeking anywhere from $10,000 to $20,000 in damages.[18]

The engineers and managers involved with the project travelled to Quebec on Sunday, September 1st, to view the wreckage and to determine the cause of the collapse. The deputy minister of the Department of Railways and Canals, as well as the former chief engineer, Collingwood Schreiber, were met by M.P. Davis, the masonry contractor, and by John Deans, Peter Szlapka, and A.B. Milliken from Phoenixville. Theodore Cooper was too ill to make the trip and sent his assistant, Bernt Berger, in his place.

Prime Minister Wilfrid Laurier and his Liberal government reacted quickly to the disaster; a Royal Commission of Inquiry made up of three prominent Canadian engineers was appointed by Laurier's cabinet the day after the collapse. Their hearings would begin on September 9th.

On Saturday morning, two days after the collapse, the coroner for Quebec, Dr. George W. Jolicoeur, empaneled a jury at the morgue in Quebec City. The fifteen bodies that had been recovered were identified and burial permits issued. The coroner and the nine-person jury, made up of prominent citizens from the area and a local member of the Quebec legislative assembly, went to the site of the collapse and spent several hours examining the wreckage. They adjourned to the following Tuesday morning, when the formal investigation into the cause of death of the workers would begin. The member of the legislative assembly was elected chairman of the jury.

CHAPTER 17

THE CORONER'S INQUEST

ON TUESDAY MORNING, SEPTEMBER 3RD, the coroner opened the inquest into the cause of death of eighteen-year-old Zephirin Lafrance, a painter working for the Phoenix Bridge Company, whose body had been recovered from the wreckage. The inquest took place in the civil trial courtroom of the Quebec City Courthouse. The jury's conclusions would be non-binding, but any recommendations they might choose to make would be given due consideration.[1]

Several lawyers were present on behalf of the parties involved, including M.M. Stuart for the Phoenix Bridge Company, L.A. Taschereau for the Quebec Bridge Company, and W.H. Davidson, representing the Iron & Steel Workers' Union.

The first witness to give testimony was the father of the deceased, whose son bore his name. Zephirin Lafrance Senior had also been employed by Phoenix to work on the bridge but had left the job two months earlier. He learned of his son's death the morning after the disaster. The boy's body had been transported to Mr. Lafrance's residence on Saturday. The young painter had been buried the previous day. Mr. Lafrance testified that his son had never expressed any fear of working on the bridge. That same day, the boy's father commenced a wrongful death action against the Phoenix Bridge Company for $15,000.[2]

The timekeeper for Phoenix was next. Adolphe Huot provided a list of the men who'd been working on the day of the collapse. He confirmed the number of workers whose bodies had been recovered, those who had been injured, and those who were still missing. The list was entered as Exhibit A. Huot told the jury that he was on the bridge at the time of the accident and he felt it start to go down at 5:31 p.m. He testified that there was no truth to the rumors that the workers feared for their safety and had refused to work on the bridge, and pointed out that the Phoenix foremen and their engineer were also at their posts at the time of the accident and that they, too, had perished.[3]

The president of the Quebec Bridge Company, S.N. Parent, was the next witness to testify. He was scheduled to attend the annual meeting of the shareholders of the Quebec Bridge and Railway Company that afternoon. Parent described the organization that had been put in place for the project. He explained the roles of the various engineers involved, focusing on the consulting engineer, Theodore Cooper. Briefly, he described the construction contracts entered into with the two contractors, Phoenix and Davis. He testified that at no time did he ever hear that the bridge was dangerous or suffered from any major defects.[4]

Parent was asked by a juror whether there was any truth to the report in the *New York Times* the previous Saturday, regarding a telegram that Cooper had allegedly sent to the site, calling the men off the work. Parent replied that upon seeing the article, he immediately sent a telegram to Cooper asking him to explain it, and that Cooper had replied that he had never sent such a telegram, nor written to the site telling the men to stop the work. Parent told the jurors that no such telegram had come from Cooper, either before or after the accident. The matter of a telegram from Cooper never came up again during the remainder of the inquest.

The last witness for the opening day was John Sterling Deans, the chief engineer of the Phoenix Bridge Company. He told the jury that the specifications for the bridge had been prepared by the Quebec Bridge Company and its consulting engineer, Theodore Cooper, and that the plans were prepared by Phoenix in accordance with those specifications. He had visited the bridge site that morning to view the wreck but was not in a position to give an opinion as to the cause of the collapse. In response to a question by the one of the jurors, Deans stated, "I am absolutely certain that the plans and specifications provided for the safe construction of the bridge, and I would not be afraid to follow the same plans and specifications for the same bridge. Every pound of steel employed was tested beforehand. We had the best picked men to work at the bridge. From the Quebec Bridge Company and from Mr. Cooper, we had the best assistance for the execution of our work. Theodore Cooper is one of the leading authorities in iron bridge construction in the States. So far as I know, the piers and shore work are absolutely intact and there is no fault with them. I never had any report or complaint against the solidity of the construction."[5]

In response to a question by Davidson, the lawyer for the union, Deans said that there had been no material changes to the original plans. He said that Phoenix never had any advice from Cooper or any other engineers that their plans were not safe enough, and added, "I am aware that every precaution was taken for the construction of the bridge."

The inquest adjourned at 1:00 p.m., to be resumed the following morning, Wednesday, September 4th.

Parent and Hoare left the courthouse and went straight to the annual meeting of the shareholders and board of directors of the Quebec Bridge and Railway Company a few blocks away. The directors of the Company had decided to open the meeting to the public and the press was in attendance. President Parent called the meeting to order, and the report that Hoare had been preparing on August 27th was read into the minutes. In a hastily prepared additional report, Hoare referred to the accident as a "national calamity," and noted that the government had appointed a commission of engineers to determine the cause of the collapse. He referred to a written statement from the prime minister of Canada, as well as a quote from the leader of the opposition, both of which stressed the importance of continuing the work. "The Quebec Bridge will be completed," wrote Hoare; "it is just a question of more time." The only encouraging news Hoare had to report was that the two masonry piers had survived the collapse largely intact. He concluded his report with a summary of the events of the last two days just prior to the collapse, following Kinloch's discovery of the bends in chord A9-L.[6]

Parent then reported on the incident involving Cooper and the article in the *New York Times*. He said that Cooper had told him that the *Times* had not reported the interview correctly, and that he would have the article corrected. This differed from what Parent had told the coroner's jury that morning.

The treasurer for the Company reported that total expenditures on the project thus far were $1.6 million for the piers and $3.1 million for the superstructure, with $164,000 still owing to Phoenix. When asked about the contract with Phoenix, Parent told the shareholders that the contract was not based on a fixed price, but on a unit price per pound of steel. Although Phoenix's original tender was $3.5

million for the entire structure, the price of steel had increased significantly, and the Quebec Bridge Company had assumed the risk of escalation.

One of the shareholders pointed out that there were now six railway lines in close proximity to the bridge, with more lines to be added. They could therefore expect a significant increase in the volume of materials to be transported over the bridge, over and above what was originally assumed. The shareholders were convinced, now more than ever, of the necessity of the bridge.

When it came time to elect the officers of the Company, Parent reminded the group that at the last annual meeting in 1906, he had expressed a desire to step down from the position of president. He had at that time been appointed president of the Transcontinental Railway, which required him to live in Ottawa. The board had insisted then that he remain as president, and he had done so. Parent told the group that in light of all that had happened, it was now more important than ever for the president to live in Quebec, and he put forward the name of someone he had discussed it with, someone from Quebec City. Following a brief debate, Parent was once again elected president. Those who elected him said that his resignation would have amounted to a second catastrophe for the bridge.

It was also proposed and accepted that Parent should be paid an annual stipend of $3,000, since the secretary of the Company received a salary and the president did not. The annual meeting of the shareholders of the Bridge Company adjourned. They were, now more than ever, determined to continue the task of bridging the St. Lawrence.

The day after the Company's meeting, a local newspaper commented upon Hoare's report regarding a deflection in one of the bridge members, two days before the collapse. The editor wrote, "This revelation must come like a shock to the public. It shows that those entrusted with the work were aware of, to say the least of it, the very questionable security of the bridge and though no 'immediate danger' was apprehended, took no precautionary measures to safeguard the precious lives that were dependent on them for protection."[7]

The inquest resumed the following morning. Edward Hoare, the chief engineer for the Quebec Bridge and Railway Company, took the stand. He testified that he had been involved with the preparation of the plans and specifications, but only up to a certain point. When asked to explain what he meant by that, he said that Cooper was the consulting engineer for the Quebec Bridge Company as well as for the federal government, and that Cooper had made changes to the general specifications in order to increase the efficiency of the structure. These changes were submitted to and approved by the federal government. Hoare stated that he himself had supervised the work from the start, and that weekly reports were sent to Cooper, who was kept up to date on the progress of the work.[8]

Asked whether he had any reports of problems with the bridge, Hoare told the jury that on Tuesday, August 27th, the young engineer Norm McLure had reported to him that chord A9-L had a slight bend in it. He sent McLure to New York and Phoenixville the next day, as the telegraph and telephone services were too slow and unreliable for discussing such an important question. He assured the jury, "I never thought that the condition of the chord was serious enough to cause an accident to the bridge. But for the reason previously stated, I thought it was of sufficient importance to send Mr. McLure to New York and Phoenixville to discuss the matter. I have no opinion to give as to the cause of the accident. I cannot say that that defect was the cause. It will take time to investigate to arrive at a positive decision. I have not the slightest doubt in my mind that all the precautions within the power of human agency have been taken by the constructors and everybody concerned with the work. The level of the bridge, the alignment of the posts, the level of the piers being absolutely correct and agreeing with calculations, showed that the bridge was in perfect condition, and that no movement whatever had taken place, or that any accident could be anticipated. I could not see any defect in the chord of the bridge from the track level because it was apparently so slight. To see the defect, it would have been necessary to stand near the chord itself." Hoare had not climbed down to see the bent chord himself; yet, in his opinion, the defect was slight.

The lawyer for the workers' union, Davidson spoke up. "You considered it a sufficiently serious enough condition to send Mr. McLure to New York to report the matter to Mr. Cooper and to consult with him about it?"[9]

"Yes, I did, principally on account of the delays of communication by phone or telegraph existing at the time, which might have caused a discussion for several days with misunderstandings, and questions of that kind in a structure of such magnitude should never be delayed."

The lawyer continued, "As an engineer, Mr. Hoare, you found a condition there in that curvature which you knew ought not to exist?"

"Yes, I did."

"Is it not a fact that the chord of which you have now spoken had only recently become curved, as you said?"

"I had two reports: one from Mr. Birks, the Phoenix Bridge Company's engineer, stating that the curvature was long standing; and one from Mr. McLure and Mr. Kinloch, that it was of recent occurrence."

"Did you measure the piece?"

"The extent of the curvature was determined by actual measurement. As an engineer, I considered that that piece should have been rectified, as it should not be allowed to remain as it was. A very precise inspection of every piece of work was made at Phoenixville and at the bridge before being put in place. I do not believe that either our own engineers or the engineers of the Phoenix Bridge Company would have allowed that chord to go in if that curvature had existed to their knowledge, and it was part of their duty to know whether it was curved or not. Supposing no external force to have been exerted upon the bridge, it is evident that the cause of the accident must be looked for in the structure itself."[10]

When he was asked for his opinion on the cause of the collapse, Hoare again replied that he would have to wait for the report from the Royal Commission engineers that had been appointed by the federal government, and with whom he was scheduled to meet that afternoon.

Norman McLure was next. He testified that on Tuesday, August 27th, he brought Hoare his measurements and a report on the state of the bent chord, A9-L. His duty was to supervise the work at the bridge, and it was the first time his attention had been called to that defect. He had not considered the defect dangerous but, to a certain extent, serious. He left at noon the following day for New York to consult with Cooper. He saw Cooper in his office at 11:30 a.m. He told the jury, "I showed the sketches I had brought with me and explained to him the condition of the said chord. Mr. Cooper appeared to be slightly worried and seemed not to be of the opinion that it was dangerous but that it should be looked after. He never told me not to put anymore load on the bridge, and he did not tell me either to telegraph any instructions to Quebec. I had instructions from Mr. Hoare to go to Phoenixville and Mr. Cooper told me also to go there. I was at Phoenixville at 4:53 p.m. and at once I went to Mr. Deans' office. He had already received a report from Mr. Birks about the same defect. Mr. Deans expressed then the opinion that that chord had always been in the same condition. I told him I was not of the same opinion. We discussed the matter a little together. He did not tell me not to load the bridge anymore. I was to see Mr. Deans again the next day, but I heard of the accident about an hour and a half after I left him."[11]

The next witness was the young Alexandre Ouimet, who had worked on the bridge since early May as a painter and laborer. At the time of the disaster, he had been standing a few feet away from the Phoenix office when he heard a noise. He glanced towards the bridge, took two or three steps, and watched in utter disbelief as the bridge collapsed. He hadn't been afraid to work on the bridge, but he told the jury about an Irishman working on the bridge who said that he wouldn't cross it for all the money in the world.[12]

Ouimet then told the jury that in May of that year, he had seen a cracked steel plate near one of the main posts. The gallery erupted. The coroner brought down his gavel and called the room to order. Hoare stood and requested that the witness be asked to identify the specific location of the plate on the plans. Ouimet was asked whether he had told anyone about the cracked plate. The witness replied no, just his three co-workers, only one of whom was still alive. The jury foreman asked the coroner to subpoena Ouimet's co-worker and anyone else who might be able to shed some light on the matter of the cracked plate. At the request of the jury, Coroner Jolicoeur adjourned the inquest at 1:30 p.m., to be re-convened the following morning.

The inquest re-convened on Thursday, September 5th, at 10:30 a.m. The coroner informed the jury that he had succeeded in finding another witness to shed light on the evidence of the previous day. The young worker, Ouimet, took the stand again in front of a packed courtroom. He was handed a photograph of the bridge taken two days before the collapse and asked to indicate to the jury where the cracked plate was on the bridge. The coroner also asked him to point it out to McLure. He did so and then Ouimet told the jury that he had seen the plate a second time in July and that the crack was still there—no repairs had been made. Asked whether he told the engineers about it, he said he hadn't because he didn't think it was dangerous, and also because he had often seen the engineers in the area and thought to himself that they must have been aware of it.[13]

Ouimet's co-worker Raoul Lafrance was then called to the stand. Raoul was Ouimet's cousin and the brother of the deceased Zephirin Lafrance. Although he had left the project ten days before the collapse, Lafrance had remembered seeing the cracked plate in July when Ouimet had shown it to him. He had seen it again in early August, fifteen days before he left the job. He was asked to describe the crack. Lafrance told the jury the crack was about twenty inches long and the width of his small finger. He, too, thought that the inspectors must have seen the crack, and so he didn't say anything about it either.[14]

The young inspector Norman McLure was then recalled. McLure told the jury that the plate was not cracked, it was crimped. He took a piece of paper and bent it to show the jury the manner in which the plate had been crimped. McLure testified that the plate had been crimped that way intentionally to allow it to be positioned into the structure. He had inspected the plate many times, as recently as August 8th, and had records to prove it. He admitted that from a distance, the crimp may have been mistaken for a crack.[15]

McClure was asked whether he thought that this plate, if it had been cracked as Ouimet had described, could have caused the collapse of the bridge.

"Not at all," McLure replied.

"Are you perfectly sure of that?"

"Perfectly sure. It is absurd to think so."

McLure explained that the plate was used to hold the end of the bottom laterals and was there solely for wind bracing; it did not bear any weight from the bridge.[16]

The Quebec Bridge inspector Kinloch then took the stand for the first time. He had heard McLure's evidence and he was very familiar with the plate. He confirmed that the plate was crimped, not cracked. He was absolutely positive that the plate was fine. Although most members of the jury seemed satisfied that the plate could not have caused the collapse of the bridge, some continued to have doubts. The inquest moved on.[17]

Kinloch told the jury that he had been on the bridge, performing his last inspection for the day, just fifteen minutes before it went down. He described what he saw when the bridge collapsed. The coroner then asked him, "You don't know, and you cannot say perhaps, where the break first started? You have no idea?"

"I have an idea."

"Where?"

"The panel 9 of the bottom chord."

"You think it started there?"

"Yes, sir."

"Had you noticed that same day that defect in the chord of panel 9?"

"Yes, sir. I looked at it when I came in."

"Did you think then that it was dangerous?"

"I would not have been on the bridge if I had."

The lawyer for the unions, Davidson, then examined Kinloch. He asked him whether he had seen the chord since the accident and what condition it was in. Kinloch said the chord was badly bent up and buckled, in the shape of the letter S. He confirmed that the bends in the collapsed chord were in the same direction in which it was bent before the collapse, and that such bends could only be caused by pressure applied at both ends of the chord. When asked whether he was aware of the bends in the chord before the collapse, he told Davidson that he had discovered the bends two days before the collapse at 9:00 a.m. on Tuesday morning, and had reported it to Yenser, Birks, and McLure. Kinloch relayed the conversation that took place, recalling Yenser's comment that the bend had not been there before, and Birks saying that the bend might have been there all along. Asked if they had considered it a serious matter, Kinloch said that all but Birks considered it to be serious.

"You had inspected, no doubt, that same part, many times?"

"Yes, sir. Probably a couple thousand times."

"On the morning of the twenty-seventh of August, about nine o'clock in the morning, was the first time that you ever discovered this curvature?"

"Yes, sir."

"If it had been there before, you would undoubtedly have seen it?"

"Yes, sir."

"Were you employed as inspector of the bridge when that particular piece was placed in position?"

"Yes, sir."

"Was it then in perfect condition?"

"Yes, sir."

"It was?"

"Yes, sir."

"Was it in all respects like other chords that were put into the bridge?"

"All except for a repair on one end."

After explaining the damage sustained by the chord when it had been dropped in the storage yard, and the repairs that had been carried out, Kinloch was asked by Davidson, "Now, Mr. Kinloch, can you explain to the jury why, having discovered such a serious condition as you did discover on that morning, you or someone else in authority did not stop the work on the bridge until an investigation could be held?"

"Well, in the first place, I had no power to stop it. It was generally supposed that we would not put up any more steel until we would be able to get engineering advice on the chord."

"And notwithstanding that general impression, the work of putting up the steel still continued during the day?"

"There was no steel put up that day at all."

"Now, on that very day or the day previous, was there any question of the advisability of not moving the traveller farther out for the present?"

"It was talked of, yes, sir. There was a general discussion amongst us at the time about it. Mr. McLure and myself did not want to move the small traveller farther out."

"Why?"

"Because I did not know enough to know what was the matter with that chord."

"You considered then, Mr. Kinloch, that it would have been prudent not to have moved that traveller farther out, in view of that curvature in the chord that you had discovered, until an investigation could be held?"

"At that time I would, yes."

"In view of what happened afterwards, Mr. Kinloch, you found that your opinion as to the prudence of not moving out the traveller as the correct one? In other words, has your former opinion now been confirmed, in view of what happened?"

"My opinion of what?"

"Your opinion was, Mr. Kinloch, as you told us, that after making that discovery of the curvature in chord number 9, an examination ought to be made, and that the traveller ought not to be moved out until it would be made?"

Kinloch didn't respond; the courtroom was still.

Davidson continued, "At any rate, Mr. Kinloch, you are now of the opinion that the break in the bridge occurred at that chord in panel number 9?"

"It occurred in two of them at the same time. It must have. Not only in one."

"And if, after the discovery was made of the curvature in the chord and the matter reported, as you have described, to the proper officers, the work had been all stopped, of course, nobody would have lost their lives?"

"If they had stopped the works, nobody would have lost their lives."

The lawyer for Phoenix, Stuart, stood up. "Mr. Birks was of the opinion that there was no change in the chord?"

"No. He thought that it might have been in there all the time."

"He did not take a serious view of the matter. At least, not as serious as you did?"

"No, not as serious as I did. But he wanted to go into it, and then Birks and I went over every lacing and every joint, and everything in it microscopically."

"Now, Mr. Kinloch," Stuart continued. "Neither you, nor Mr. Birks, nor any other person anticipated that catastrophe, or you would not have been on the bridge?"

"No, sir."

"And all that you saw was that it was sufficiently serious to require the careful attention of the engineers?"

"Yes, sir."

"And, in fact, the following morning, after Mr. McLure had reported to Mr. Hoare, Mr. Hoare went up to visit the chord in question?"

"Yes, sir."

"And Mr. McLure went down to New York, as was said before?"

"Yes, sir."

McLure was excused and Hoare was re-called. Juryman Gale asked him, "Who had the authority to stop the work?"

Hoare replied, "I could have stopped the work by communicating…" He paused, had second thoughts, and started again. "The proper authority, I suppose, would have been the chief engineer of the Phoenix Bridge Company, Mr. Deans."

Another juror asked him, "At the bridge, could anybody stop the works?"

"No, not anybody."

Juryman Gale asked him, "Then nobody had authority to stop the work, only the Phoenix Bridge Company?"

"Direct orders to stop the work should go through them, either through their chief engineer at Phoenixville, or the superintendent of construction, or the general foreman of the work."

The coroner stepped in. "But, in fact, anybody could have had the right to stop the work if they had noticed the danger and called off the men."

Hoare did not answer. Again, the room was hushed.

Stuart, the lawyer for Phoenix, broke the silence. "If there had been any danger anticipated, you would have told the engineer at the bridge, and the men would have been called off?"

Relieved, Hoare replied, "If I had been aware of it, I would have given notice to the foreman of the works, but my duty would have ended there. I don't say the men would have been called off. I would have notified the foreman of the works." Hoare hadn't heard the last of this line of questioning.[18]

Kinloch was recalled to clarify some of his previous evidence. The coroner asked him, "Mr. Kinloch, after all that you have said to Mr. Davidson and to the jury, are you of the opinion, or were you of the opinion, at the time at which you noticed that curvature in chord 9, that it offered immediate danger for the bridge?"[19]

"No, sir."

"If you had been the first authority there that day, would you have ordered everybody off the bridge that day? Did you think the matter serious enough for that, or did you think that the bending in that chord was serious enough to make it dangerous?"

"I thought it was serious. I was on the bridge myself after that."

"You did not think it was serious enough to bring the collapse of the bridge?"

"No, I did not think anything at all about that. I did not stop to think of anything of that kind."

"Now, do you think it would have been wiser at the time to call everybody off the bridge and stop the work?"

"Undoubtedly, at the time."

"Did you suggest it at the time?"

"What? Take the men off?"

"Yes."

"No, sir."

Juryman Gale asked him, "If you had had the authority to stop the work, would you have stopped it?"

"I would not have gone ahead until I found out what was the matter with that chord. I would have stopped it until I found out."

Stuart, the lawyer for Phoenix, stood and asked him, "What did you mean by suspending the work?"

"I would not have put on any more steel."

"And in fact, you told us that instead of more steel being put on, some steel was taken down?"

"Yes, that day."

"Now, by the answer you gave to Juryman Gale, you did not mean to say that you would have ordered all the men off the bridge, simply that you would not have increased the weight until you had an opportunity to investigate the cause of the bend in the chord?"

"Yes, sir."

Davidson, the lawyer for the workers' unions, chimed in. "You said a moment ago that the pressure or the weight was lightened that day. Was it added the following day?"

"It was."

"That is, during the day the accident happened, more steel was put in position on the bridge?"

"Yes, sir."

"And you, as Inspector, did not consider that that was not a wise or prudent thing to do in view of what you had discovered?"

"At that time, I had figures, and it was all right."

"From whom?"

"From Mr. Birks."

"And Mr. Birks was the Phoenix Bridge Company's engineer?"

"Yes, sir."

"And you accepted his statement that it was all right to go ahead?"

"Yes, sir."

"Well, now, Mr. Kinloch, as a last question, and to come back to one which has been asked you a good many times, I will put it in good English, which you can readily understand, and I hope you will give me as plain an answer. After having made the discovery which you did, as to the chord being bent or curved, and in view of the seriousness of the matter as it appeared to you, if you had had supreme authority, would you have stopped all work on the bridge until the matter was investigated? Now, I don't think I can make it any plainer."

Kinloch replied, "I would not have put any more steel. I would not have stopped all work because I would not have figured that it would make it any worse."

Davidson shook his head and sat down.

The coroner advised the jury that there were no other witnesses to be called that day. The foreman asked the coroner whether he thought there would be more witnesses in the future. The coroner told the jury that he could not be sure whether there would be or not. The foreman looked at his fellow jurors and then turned to the coroner. "The jury's duty," Juryman Gale said, "is to discover the cause of death, which was a simple task in this case, but also to determine who might be criminally responsible for the death. This responsibility," he pointed out, "might be more easily determined by the inquiry to be conducted by the three engineers recently appointed by the federal government. We are convinced that there are indeed people who can shed light on the subject matter of the inquest, and we ask these individuals to come forward to help bring to justice those who were responsible for the disaster."

On behalf of the jury, the foreman asked that the inquest be adjourned sine die. Coroner Jolicoeur agreed and adjourned the inquest without setting a date when it might be re-convened.

CHAPTER 18

THE ROYAL COMMISSION OF INQUIRY

THE THREE COMMISSIONERS APPOINTED BY Laurier's government were instructed to "conduct an investigation into the cause of the collapse of the Quebec Bridge in the course of construction over the St. Lawrence River, and into all matters incidental thereto." They were to report the results of their investigation to the federal government, and to "provide any opinions they saw fit to express." Two of the commissioners travelled to Quebec City on Friday, the day after the collapse, and the third arrived on September 4th. They received their formal commission on September 9th and began taking evidence that afternoon in the criminal trial courtroom of the Quebec City courthouse. The commissioners had the power to summon witnesses to testify under oath, and to order them to produce all relevant documents in their possession.[1]

The chairman, Henry Holgate, was a forty-four-year-old engineer who had been involved with railway and bridge construction his entire career. He had been apprenticed to Colonel Fred Cumberland in the construction of the Northern Railway of Canada in Ontario. After sixteen years in railway construction and bridge design, he moved to Montreal where he worked in the construction and operation of electric railways. Holgate was highly regarded by his peers and was a member of both the Canadian and American Societies of Civil Engineers.[2]

John Galbraith, the senior commissioner at age sixty-one, was a well-known Toronto engineer and educator. Early in his career, Galbraith had worked for the

Midland Railway in Ontario and then he did surveys for the Canadian Pacific's main line. He obtained a master's degree at the University of Toronto, eventually becoming chairman of engineering in the university's new School of Practical Science. Twelve years later, Galbraith was appointed dean of the newly formed Faculty of Applied Science and Engineering. Shortly after his appointment to the Commission, he was elected president of the Canadian Society of Civil Engineers.[3] In 1908, H.E.T. Haultain would become a professor of Mining Engineering at the University of Toronto, and eventually head of the department. Galbraith and Haultain were colleagues.

John Kerry, the youngest member of the panel at age thirty-nine, was a gold medalist from McGill University, graduating with a degree in civil engineering. Kerry had worked in railway survey and engineering prior to joining the faculty of applied science at McGill in 1896. Just prior to his appointment to the Commission, he had left academia to pursue a private practice in Toronto.

A fourth engineer, who was not a member of the Commission, was hired separately by the Department of Railways and Canals to provide an independent opinion on the design of the Quebec Bridge. Charles Conrad Schneider was a well-known bridge engineer from Philadelphia. Among his numerous accomplishments, Schneider designed the interior steel framework that supports the Statue of Liberty. In fact, it was Schneider who Cooper had recommended to take his place in 1905, when it became evident that he would be unable to travel to site. Schneider would eventually find the design woefully inadequate.

The inquiry began on Monday afternoon, September 9th. The courtroom was filled with families and friends of the victims and witnesses summoned to testify, as well as reporters and members of the public. Only eleven days had passed since the collapse. The loss of life was staggering, one of the worst industrial accidents in Canadian history. Every person with a connection to the disaster was still in a state of shock and looking for answers. Many families had thus far been denied closure—less than half the bodies of those killed had been recovered. The commissioners had a difficult task ahead of them—to discover the truth, determine the cause, and hopefully, find out who was responsible for the terrible loss of life.

The coroner's inquest had begun the week before in another courtroom just down the hall from where the commissioners sat, and the testimony before the coroner's jury had been reported daily in the local newspapers. Although the coroner and the jury had thus far been unable to determine the cause of the collapse, a number of issues had been raised by the witnesses, leading to much speculation. Everyone now looked to the three engineers on the dais to shed more light on the cause of the collapse.

Conspicuous by their absence in the courtroom were the two design engineers, Peter Szlapka and Theodore Cooper. Cooper advised the Commission that his doctor had forbidden him to travel; if the commissioners wished to examine him, they would have to travel to New York. Szlapka would be examined in Phoenixville. Neither men would ever face the survivors or the families of the victims. Cooper would never see the wreckage for himself—he never saw the superstructure while it was being erected, or after it collapsed. He saw only photos.

The president of the Quebec Bridge and Railway Company, and chairman of the board, S.N. Parent, who had also been the mayor of Quebec City and the premier of the province of Quebec until 1905, also did not appear before the commissioners. He, too, did not have to face those affected by the tragedy. He made a donation of fifty dollars to the families' relief fund. Each of the commissioners also donated fifty dollars to the fund. Parent's evidence consisted of a written response to questions raised by the commissioners.

Chairman Holgate opened the hearing by reading the Royal Commission granting them authority. He then told those present that the commissioners would conduct the examination of witnesses themselves, but would nevertheless welcome suggestions, particularly from the lawyers who were present. There were four lawyers in attendance, three of whom had attended the inquest. The fourth lawyer, who had not attended the inquest, was the general counsel for Phoenix. Phoenix had been represented by a local Canadian lawyer from Quebec City, Gustavus G. Stuart. The Quebec Bridge and Railway Company's outside counsel, Roy, was present, as well as Davidson, the lawyer for the two unions representing the workers, the International Association of Bridge Workers, and the Bridge and Structural Iron Workers' Union.

Without being invited to speak, John Hampton Barnes, the Phoenix lawyer from Philadelphia, stood. "Mr. Chairman, I wish to advise on behalf of my employer that it tenders its fullest and heartiest co-operation to the Commission. We will therefore, sir, be ready at all times to furnish documents, records and plans which are not available by the exercise of your subpoena in this jurisdiction, and which are in our control and possession outside of this jurisdiction. We will also be subject to your directions in the production of the officers, representatives, and employees who have knowledge of the facts."[4]

Holgate replied, "Thank you very much, Mr. Barnes; I quite felt that that disposition would exist on the part of the Phoenix Bridge Company."

Counsel for the unions, Davidson, then stood. "I just wish to add that I heard with a great deal of pleasure the words that have just fallen from the lips of the learned counsel of the Phoenix Bridge Company."[5]

After the swearing in of two Hansard stenographers, the first witness called to testify was the secretary of the Quebec Bridge and Railway Company, Ulrich Barthe. Barthe was the Quebec City councillor who had been persuaded to act as secretary of the Company by the president of the Quebec Bridge and Railway Company and mayor of Quebec City, S.N. Parent.

Holgate asked Barthe for the contract documents, the plans and specifications, and the organization chart, as well as the contract between the Company and its consulting engineer. Barthe had come unprepared; he was given one day to gather up the documents.

Holgate's examination of Barthe continued, focusing on identifying who the chief engineer for the project was. Barthe testified that E.A Hoare was the chief engineer, and that Theodore Cooper was the consulting engineer. When asked whether there was a resolution from the board giving either of these engineers precedence, in any way, over the other, Barthe replied, "These are not questions which I would like to answer." Holgate pressed the issue and Barthe finally admitted there was no resolution giving either of them precedence over the other; although there were resolutions of the board appointing them, their respective duties were not defined. Barthe was dismissed and ordered to return the next day with the requested documents.[6]

The second and last witness on the first day was John Sterling Deans, the chief engineer for the Phoenix Bridge Company. Holgate had Deans take the Commission through the complete Phoenix organization, including the design office, fabrication facilities, inspection regime, and the organization put in place for the erection of the bridge. Deans told the Commission that Yenser was the man responsible for the erection of the structure, but that he acted under the advice of the engineers. Although Yenser was the final man in authority, he would not have done anything contrary to the instructions of the engineers, and in fact, the instructions on the plans and blueprints could not be departed from without instructions from the Phoenix office. Asked whether this method of proceeding was normal, Deans answered, "That is a very unusual method. It was adopted in this case as an extra safeguard against allowing the foreman to use his judgment in regard to the handling of the material, and to fix that method by the very best experience we had in our company."[7]

The Commission adjourned, to reconvene the next day at 10:00 a.m.

Barthe appeared on Tuesday and regretted to advise the Commission that he had been unable to gather all of the information requested. He was given another day.

Deans was recalled and asked who had the authority to dismiss Yenser or Birks. He replied, "Mr. Milliken could have dismissed Mr. Yenser, and could have requested the removal of Mr. Birks from me, and his request would have been conceded." Deans told the Commission that although Cooper did not approve the erection plans, they were sent to him for information, and that "Mr. Cooper, being in supreme authority, could have stopped or interfered with the erection through Mr. Hoare at any time that he saw fit."

Professor Galbraith then asked him, "When you said yesterday that these plans were simply a part of the Phoenix Company's business, and that Mr. Cooper had no responsibility in connection therewith, you meant that to be simply in general, while at the same time, as you have said now, he could stop the work, and stop anything that he pleased, through Mr. Hoare?"[8]

"Absolutely."

"That is the position?"

"That is the position."

Holgate asked Deans to explain Hoare's authority and position in relation to the contract. Deans replied, "Mr. Hoare being the chief engineer of the Quebec Bridge and Railway Company, I understood that I should look to him for any final instructions in connection with the contract in its execution."

"Of what nature?" Holgate asked.

"Well, I should say in all matters outside of the approval of plans and the interpretation of the specifications, and should look to him for final instructions in connection with the work in the field or shop."

Professor Galbraith chimed in. "In other words, you assumed that he had the power to stop any piece of work?"

"That expresses it," Deans replied.

Galbraith emphasized, "To reject any piece of work."

"That expresses it."

Professor Kerry got straight to the point. "Or, to express it in another way, your understanding would be that—with the exception of the preparation of the specifications and the approval of the detail plans—the entire responsibility for the construction of the bridge lay with Mr. Hoare?"

"I do not think that is expressing it too broadly," Deans answered.

Galbraith moved on. "Mr. Deans, if Mr. Cooper should have given instructions to stop the work at any stage, what would you have considered it your duty to do?"

"I should have felt it incumbent to notify Mr. Hoare and receive his instructions."

All eyes were on Hoare.

On the afternoon of the third day, Edward A. Hoare, the chief engineer for the Quebec Bridge and Railway Company, took the stand. He would be called to testify eleven times. Holgate began by asking him to describe his background in bridge design and construction, and the organizational structure established for the project.[9]

With the preliminaries out of the way, Holgate asked Hoare whether the Quebec Company's chief inspector, McLure, had the power to dismiss an employee of Phoenix; Hoare told him no. Holgate then asked whether Hoare had that power.

"No, sir."

Holgate paused. "Was there any power vested in Mr. Cooper?"

"No."

"Had any of the officers of the Quebec Bridge Company power to stop the work of the Phoenix Bridge company?"

"Any of the officers?" Hoare asked.

"Either Mr. Cooper, yourself, McLure, or Kinloch?"

"To stop the work?"

"To stop the work of the Phoenix Bridge Company?"

"I do not know that there is anything in the contract which would give us any such power," replied Hoare.

Holgate and his fellow commissioners were stunned.

Holgate gave Hoare an out. "Will you say now that, at the present time, you cannot answer that?"

Hoare was wary but failed to appreciate the gravity of the question. He asked the chairman, "Will you give me a minute? I will just see."

Holgate warned him, "It is a question that requires very careful consideration."

Hoare ignored the warning; he was determined to answer. "I want to read over the contract to see whether there is any power vested in the contract."

The commissioners waited patiently as Hoare read over the ten-page contract. Finally, Hoare declared, "There is nothing in the contract."

Holgate couldn't believe what he was hearing. "What is your understanding?"

"We have got no power in the contract."

"After reading the contract now, you conclude you have no power?"

"There is no power in the contract itself, no clause in the contract giving anybody connected with the Company power to stop the work."

Holgate was speechless. Professor Kerry spoke up. "Had you considered previously, Mr. Hoare, whether you had such power?"

"No, I never considered the question at all."

Holgate asked him, "Then, do I understand that this is the first time you have considered that question?"

"Yes." Hoare had forgotten that he'd been asked the same question at the inquest.

The courtroom buzzed. Hoare was dismissed and ordered to return the next day. The Commission adjourned for the day. They needed time to digest Hoare's evidence.

The next morning, Thursday, September 12th, the hallways of the Quebec City courthouse teemed with people. The coroner had re-convened the inquest into the cause of death of the young Zephirin Lafrance, and down the hall, in the criminal trial courtroom, the Royal Commission of Inquiry was entering into its fourth day of hearings. Hoare had testified the previous day that neither he nor anyone else from the Quebec Bridge Company had the power to stop the work.

Hoare was recalled to the stand. Holgate asked him, "Mr. Hoare, will you recall the last part of your examination of yesterday? Have you any explanation you would like to make in connection with it? That referred to your power as chief engineer of the Quebec Bridge Company; you gave us the impression yesterday that you had no power to stop the work of the Phoenix Bridge Company, and you told us before that, that you could not dismiss any men in the employ of the Phoenix Bridge Company?" Holgate was giving the man one more chance.[10]

Hoare had indeed given considerable thought to his response of the previous day. He told the commissioners, "I would like that statement in reference to the question of stopping the work qualified, and the following statement substituted: Notwithstanding that the contract does not refer to any power vested in the engineers for stopping the work at any time, I can say that if any serious question arose affecting the structure, or if there was serious damage to any part of the structure, under such circumstances I would stop the work."

The damage was done, however. Hoare's uncertainty as to his authority, as the owner's representative, to stop the work, pointed to a weak, dysfunctional organization clearly unable to deal with emergencies. Hoare had also forgotten the incident in 1905 when he had stopped a portion of the work, at Cooper's request.[11]

Holgate then described what the commissioners viewed as the chief engineer's authority. "Of course, we have our own ideas with regard to the powers and duties of a chief engineer on works, and our interpretation of that title generally would be that the chief engineer was the absolute authority on that work, that he would have the power to reject material if he did not approve of it, that he would have power to dismiss any employee of the contractors that he considered was incompetent, or was doing work improperly, or was misbehaving himself on the work, and that he would have power over the whole work to the extent of stopping any portion of the work during its progress, or the whole of the work, if, in his opinion, it was not being carried on entirely to his satisfaction, having in mind the letter and the spirit of the contract, the specifications and plans."

He then asked Hoare, "Now, with that definition, would your position correspond with its duties?"

Hoare replied meekly, "Yes, sir."[12]

Holgate moved on. "Then who was primarily responsible for the specifications?"

"The original?"

"The specifications under which the work was carried out."

"I was primarily responsible for the original, and, for the modifications, Mr. Cooper— responsible for the changes."

The young commissioner, Professor Kerry, asked him, "Was any reference to you necessary on the part of Mr. Cooper, or was Mr. Cooper the absolute authority?"

"He had absolute authority to deal with the question."

"In that respect, then, Mr. Cooper was the chief engineer of the bridge?"

"No, sir, he was the consulting engineer, and his appointment as consulting engineer gave him power to make changes."

"That is to say, his appointment delegated part of the authority of the chief engineer to him?"

"That is right."

Holgate then summarized Hoare's knowledge and experience. "We understand you to be an engineer of general knowledge, and that your professional work has led you through a varied and rather broad experience in construction and design."

Hoare agreed.

Holgate continued, "But there has not been placed before us any evidence that you assume to be an expert or an authority on bridge construction as a specialty."

"No."[13]

"Now, will you tell us, was the appointment of a specially qualified bridge engineer, a man who would have experience in the erection of large bridge structures, ever discussed? Someone whose duty it would have been to remain on the ground during the construction of this work—was the appointment of such a man ever discussed?"

"Not to my knowledge." Hoare testified that he didn't think it was needed, nor did he recommend such a thing. He told Holgate, "Mr. Cooper was informed, sometimes daily, and always at the end of each week, of the daily progress of the work, and was always consulted on any question of importance that arose from time to time."

"But Mr. Cooper was not on the ground?" Holgate replied.

"No, he was in New York."

Professor Kerry stepped in. "Then we would understand, Mr. Hoare, that on a very great and necessarily dangerous work, the Quebec Bridge Company was relying for its direction on a fully qualified man who could be described as permanently resident in New York, and that the only evidence he had to guide him were the reports of Mr. McLure."

Hoare asked him to repeat the question; the stenographer read the question.

Hoare responded, "I do not know that the word 'dangerous'—I do not know whether the work could be considered dangerous."

Kerry would have none of it; seventy-five men were dead. "The answer to the general question, Mr. Hoare; is it a correct statement of facts? Kindly tell us if that is a correct statement of facts?"

"Yes, that is all right," was Hoare's weak reply.

The witness was excused.

THE ROYAL COMMISSION OF INQUIRY

The timekeeper, Joseph Huot, took the stand and provided the names and occupations of the men who had been working that day, the names of those injured, the men still missing, and the identities of the bodies recovered. He confirmed that the bridge collapsed at 5:31 p.m. and told the Commission where he was when it collapsed, and what he could remember of it. "To say the truth, I saw very many things, but I cannot very well describe what I did see because I realized in a second I was in danger, and I had to escape, and I made the best I could to escape myself. I turned around and jumped, and I had to run up the hill to make the approach span."

The witness was excused at 12:25 p.m.

As Huot returned to his seat, Hoare stood and approached the dais. He asked the commissioners if he could clarify some of his answers. The three engineers discussed it among themselves for several minutes, and finally agreed. Hoare took the stand and Holgate asked him, "Prior to the collapse of the Quebec Bridge, had anything abnormal or unexpected occurred during the construction which, in your opinion, required the special attention of an engineer of special qualifications, as a bridge engineer?"[14]

"My answer I have written out as follows," Hoare replied. "I may say that the work of erection followed an entirely normal course. The tests made showed the deflection expected occurred, and the whole construction up to the time of the collapse followed the anticipated course. I was myself frequently on the works and it never occurred to me that, with my long experience, I was not absolutely qualified to superintend the construction of the bridge, and I still think so. If anything abnormal had occurred I should have sent for Mr. Cooper, but nothing suggesting the slightest danger to the bridge occurred, and I do not now see what difference Mr. Cooper's presence here during construction would have made."

Holgate asked him, "May not, in a work of this nature, Mr. Hoare, abnormal or unexpected conditions arise at any moment? I am speaking of a structure of this nature. May not they arise at any moment?"

"It is quite possible," Hoare replied.

"Is it not a thing that you might almost expect?"

"No, I would not say that."

"At any rate, if you would not go so far as to expect them, would you not prepare for them?"

"I consider that we prepared for them."

"Then when you did send to Mr. Cooper, you considered the question abnormal?"

"No, I would not say that. We made a practice of keeping Mr. Cooper thoroughly posted on everything that occurred from day to day."

The three commissioners just sat there looking at Hoare. He appeared restless, as if he had something more to say.

Holgate asked him, "Is there anything you want to say?"

"No, I do not think so. It was only to make that a little more clear. It left it in rather an indefinite position."[15]

The inquiry adjourned for lunch. Hoare and others wandered down the hall of the courthouse to find out what had taken place at the inquest that morning.

CHAPTER 19

THE INQUEST ENDS

CORONER JOLICOEUR HAD RE-OPENED THE inquest at 10:15 a.m. on Thursday, September 12th, in the civil trial courtroom. One of the nine jurors was ill and unable to attend. The coroner and his jury had decided to proceed with the inquest without waiting for the Royal Commission to complete its work. There were three lawyers present. Phoenix was represented by another lawyer from Quebec City; the lawyer for the Quebec Bridge Company, Mr. Roy, had chosen to attend the inquest and not the inquiry; and another lawyer represented the Canadian government. Mr. Davidson, the lawyer for the workers' unions, was attending the Commission of Inquiry; there was no lawyer representing the workers.

The coroner called the packed courtroom to order. Augustus Milliken, the fifty-year-old general superintendent of erection for the Phoenix Bridge Company, took the stand for the first time. He was asked to explain to the jury what precautions were taken in the construction of the bridge. Milliken told the jury that plans for the erection of the bridge had begun in Phoenixville more than two years before any erection work had started in the field. These plans had been prepared by the engineers in his department in Phoenixville. The plans were then sent to the office of the chief designer for review and approval, and they were finally sent to the general foreman of erection at the site and fully carried out. He confirmed that all pieces for the bridge were examined before they went into the structure, and that the erection of the bridge members themselves was carried out in accordance with detailed instructions from the engineers in Phoenixville.[1]

Milliken was asked about chord A9-L. He told the jury that he had left the site on Monday, August 26th, before the bend was discovered. As far as Milliken knew, the chord had left Phoenixville in good condition. He was aware of the accident with the chord in 1905, and the repairs that had been carried out in the yard. The repairs were made in accordance with instructions from the engineers in Phoenixville, as approved by Cooper. He understood that once the piece was repaired, it was as good as it was before the accident.

"This piece number 9," asked one of the jurors, "when it started from Phoenixville, was it in good condition, in perfect order?"

"As far as I know, yes, sir," Milliken answered. "But I did not… I saw very little of the material at Phoenixville. In fact, I had nothing to do with it until it reached the field."

"You did not get any report that there was an accident to this number 9 piece, in the yard?"

"Yes, there was one chord section—I knew of a chord section that had been dropped. I knew of its repair in 1905, I think."

"How was this repaired, do you know? Was it hammered or straightened?"

"No, it was repaired in accordance with instructions from the computing department, as I understand, approved by Mr. Cooper."

"When this piece was repaired, do you think that this piece was as good as before it was repaired?"

"Well, I have every reason to believe it was, if Mr. Szlapka and Mr. Cooper said it was, because we were working right practically under what they said was all right before that."

The jury foreman, Gale, asked Milliken, "Are all the pieces tested before they leave the factory?"

"All of the heats. As I understand, the bars are tested after they are rolled into bars, shapes, different shapes, angles."

"Is it your own idea that that is the piece which started first to come?"

"No, it is not. I have not reached that point where I can form any idea as to the real cause of the accident."

The coroner then asked him, "Do you think it probable by what you know about the erection, if it could come from that place, the breaking of the bridge, from that chord number 9, considering where it is situated?"

"Well, if there was no chord section on there, the bridge would likely collapse; there is no question about that."

"Yes, but there was one?"

"Yes, sir."

Roy, the lawyer for the Quebec Bridge Company, asked Milliken, "Who was in charge of the work in Quebec?"[2]

"Mr. Yenser was in general charge of the work, whether I was here or absent."

"And Mr. Yenser's duty was to report anything defective that could be found out during the erection of the work, to you or to the head office?"

"If I was here, he would confer with me first of all."

"At the time of the collapse of the bridge, Mr. Yenser was alone in charge of the bridge, of the erection work?"

"Yes, sir, at the time of the collapse of the bridge, he was in general charge."

"Now that member, that chord number 9, was put into place in 1905, you say?"

"Yes, sir."

"And anything defective about that chord was found out since 1905?"

"No, sir," Milliken replied.

"Up to the present time, up to the time of the accident?"

"Not to my knowledge, nothing repaired."

Milliken made no reference to the bends discovered by Kinloch on August 27th. Roy did not pursue it, and neither did the coroner or the jury. The witness was excused.

Norman McLure, the Quebec Bridge Company's twenty-seven-year-old inspector, was the next and final witness to be called. He was examined by the coroner.[3]

McLure testified that at the time of the accident, chord A9-L was carrying seventy four percent of the load that it was designed to carry.

"How do you account for the bend coming into that chord number 9?"

"I have not had time to account for it yet."

"Do you think it was due to the compression of the chord? Could it be compressed enough to be bent that way?"

"That is the question. Theoretically it could not, if it was straight in the first place."

"So you mean, if that chord had been straight in the first place it could not have been bent; is that what you mean?"

"Not according to the figures. Practically, it might be an entirely different matter."

The coroner moved on. "You left on Wednesday for New York yourself, the day after the defect in chord 9 was discovered?"

"Yes, sir."

"That day, the 27th of August, were you there when the measurements were taken of the alignment of the posts, and the track level?"

"Part of the time."

"And everything was correct?"

"I was satisfied that it was."

McLure was excused and the coroner advised the jury that there were no further witnesses to be called. A discussion then took place among the jurors regarding the young engineer's testimony. At noon, the jury retired.

Everyone had assumed that the deliberations would last for the remainder of the day, and perhaps go into the next day, but at 12:45 p.m., the coroner invited the reporters into the jury chamber. The foreman announced that they had reached a unanimous verdict.

Juryman Gale stood and addressed the coroner. "Mr. Coroner, we, the jury, find that the deceased, Zephirin Lafrance, died on August 29, 1907, in the parish of St. Romuald, County of Lévis, District of Quebec, from injuries and nervous shock suffered from the collapse of the Quebec Bridge. We have been unable to determine the cause or causes of the collapse of the bridge, but we feel it is our duty to declare that, based upon the evidence heard during the inquest, all precautions were taken to ensure the construction of the bridge without danger."[4]

The verdict applied equally to all victims of the disaster. The matter of criminal liability appeared to be settled—no one was held responsible for the loss of life. Hoare and the others could now breathe a sigh of relief.[5]

None of the reporters asked the jury why they had decided to render a verdict before all of the evidence had been presented to the Royal Commission of Inquiry. Instead, they rushed out of the courthouse to make the deadline for the afternoon edition of their papers.

The next day, *The Boston Herald* had this to say about the verdict: "It was too much to expect that the coroner's jury would be able to establish responsibility for the collapse of the Quebec Bridge. The jury was simply not able to provide any explanation for the accident, and from what it could discern from the evidence, all necessary precautions were taken to ensure that the bridge would be built without any serious mishaps. Such decisions are typical of many coroners' inquests. The cause of this terrible accident remains a mystery, which should be solved, to a certain extent at least, by the engineering fraternity. It is a task for the engineers to undertake, for the honour of their profession."[6]

The *Gazette* (Montreal) observed, "The coroner's jury has found that all necessary precautions were taken to secure the safety of the Quebec Bridge. The fact that the bridge fell indicates that something essential was omitted. Coroner's juries' findings are not always the most logical of deliverances."[7]

The *Globe* (Toronto) was more sanguine; "The verdict of the Coroner's jury was presented to-day, and imputing blame to no one was not unexpected, as not sufficient evidence had been heard to fix responsibility. The Government Commission, however, intend to go fully into the matter."[8]

The following week, *Engineering News* had this to say about the verdict: "The coroner's inquest, which had been stopped when the government commission began work, in order to obtain the benefit of the commission's conclusions, was resumed last week (although the commission has not heard more than a portion of the evidence) and the jury rendered a verdict. They found that all work had been done with the greatest possible care."

It is a mystery why the inquest was ended so early, and how such a clear verdict exonerating everyone could have been delivered by the jury.

CHAPTER 20

THE INQUIRY CONTINUES

NEWS OF THE VERDICT SPREAD quickly. When the Commission resumed its inquiry at 2:15 p.m., the courtroom buzzed. Holgate brought down his gavel. The timekeeper, Joseph Huot, took the stand once more. The chairman asked him to identify, as closely as possible, the location of the survivors who were working on the superstructure at the time of the accident. Huot put red crosses on a drawing of the bridge to indicate their location.

For the remainder of the fourth day, and for the next three days, the Commission examined the survivors who were physically able to appear before them at the courthouse. The workers described what they had experienced at the moment of the collapse, and during the fifteen seconds it took for the bridge to fall.

One of the survivors, Haley, the union boss, had seen the bent chords in the cantilever arm himself and had even taken measurements with a co-worker. Haley had caused a stir among the men. Holgate put it to him that he himself didn't consider these bends to be of that great importance, and the fact that he had gone to work suggested that. Haley replied, "Well, it fooled us; we did not think it would go so quick, that is all."

Holgate continued, "But you did not think them so serious as to keep you from working on the bridge."

"I thought them very serious, but I thought, surely I would have a chance to look at them the next night."

Holgate, at least, seemed to be convinced that no one believed there was imminent danger, not even the workers' representative.[1]

On Saturday, Horace Clark, the foreman in the Belair storage yard, took the stand. He described the incident in the yard when chord A9-L had fallen and the repairs that had been made to it. He pointed out that the end of the chord that had been damaged, the end that first hit the ground when the chord was dropped, was still visible among the wreckage of the bridge, and that the repairs were still intact. Clark was certain that the chord had been straight when it went into the bridge, and that it had been inspected by Kinloch and Hoare after the repairs were carried out.[2]

Clark told the Commission of his last conversation with Birks, just fifteen minutes before the collapse, how the young engineer had insisted that the chord had already been bent when it went into the structure. He was asked whether Birks had been there at the time the chord was repaired, before it went into the bridge. Clark wasn't sure, so Holgate asked Deans, who was in the courtroom, to confirm whether Birks was or was not. Deans replied that Hudson, Birks' predecessor, had been the inspector at that time, and was at site when the chord was put into the bridge in 1905. Birks had, in fact, not seen the chord before or after it had been repaired, or prior to it going into the bridge.

Galbraith then asked Clark, "If you had seen a deflection of an inch and five-eighths in the chord, do you think you would have noticed it and been sure of it?"

"I think I would," Clark replied. "An inch and five-eighths is quite a bit." In fact, the deflection, as measured by McLure and Kinloch, was two and a quarter inches.

The Commission reconvened on Monday, September 16th, for their seventh day of hearings. The two workers who testified at the inquest two weeks earlier that they had seen a cracked plate had been summoned to appear before the Commission. Holgate called out their names, but Alexandre Ouimet and Raoul Lafrance were not in the courtroom. The chairman ordered the bailiff to find them and bring them to the courthouse. The Commission proceeded with the examination of other workers.

The bailiff had little difficulty locating Lafrance. He was transported to the courthouse and sworn in at 2:00 p.m. Raoul Lafrance was the brother of the young Zephirin Lafrance, whose death had been the subject of the inquest. He told the Commission that he had quit his job as a painter with Phoenix just ten days before the collapse. His foreman was his cousin, Alexandre Ouimet. Holgate asked him, "Is he here?"[3]

"No, he is gone."

"Where?"

"I do not know where he's gone to. He's left for Ontario."

"When did he leave?"

"Saturday afternoon."

"Did he tell you where he was going?"

"I don't know. I know he took the C.P.R. train and left for Ontario. He's gone to the shanties there. He gave me the name of the place, but I don't remember."

"Who knows?"

"I don't know."

"And you don't know where he's to be found now?"

"No."

Holgate, clearly frustrated, asked Lafrance about the cracked plate. The young worker gave the Commission the same information he had given at the coroner's inquest. Holgate told Lafrance, "We want you—we order you, to go on the ground tomorrow, and in company with Mr. Kinloch and Mr. McLure, endeavour to find that plate."

"Yes."

"We want you to stay until you find that plate or make sure of something and appear here on Friday morning and resume your evidence."

"Yes."

"You are still under subpoena and it is compulsory for you to be here."

"Yes."

The next witness was A.B. Milliken, the superintendent for Phoenix. He described his movements during the last few days prior to the bridge collapse. Kerry asked him, "You recall Mr. Clark's evidence of Saturday?"[4]

"Yes, sir."

"Would you have permitted that repaired chord to have been sent down to the bridge unless it was straight?"

"No, sir," the superintendent replied.

Kerry continued, "At the time of the accident, was the condition of progress better than your expectation, or otherwise?"

"Well, it was about up to our expectation, though Yenser had lost a good deal of time on account of bad weather this season."

"So that he would be a little more than usually anxious to get the material up?"

"Not necessarily so—no, sir."

Davidson, the lawyer for the unions, piped up. "I do not know if Mr. Milliken is aware of this, but in connection with the question, I may say that my information is that the Phoenix officials were being continually urged, almost to the extreme limit, by the Quebec Bridge people to push the work forward this season."

Deans shouted from the gallery, "I wish to deny that absolutely."

"I do not say that all the information I get is absolutely correct," Davidson replied.

"That is absolutely wrong," Deans retorted.

"But it does happen that a good deal I have had is correct so far."

"Not all of it—not all you expected to be."

Holgate stepped in. "Do not argue about its correctness now; we will have Mr. Hoare explain that later."

Hoare was never asked about it.

Milliken testified that they had intended to increase the workforce by fifteen or twenty-five men and that was the reason for his trip to Phoenixville on the Monday before the collapse. He had sent a bridge man out to secure men and he had already done so near Boston. He had them under orders to report to Yenser in Quebec. Phoenix had received their bridge man's telegram on Friday morning after the accident and had wired him to hold the men and report to Quebec.

On the eighth day, September 17th, the Commissioners visited Delphis Lajeunesse, who was convalescing at his home in St. Joseph de Lévis. They then proceeded to the Lévis hospital to examine Alexander Beauvais. Holgate then interviewed Oscar Laberge at his residence near the bridge. Laberge had suffered serious injuries, including a broken jaw and fractured pelvis and right leg. The twenty-three-year-old died shortly after his examination by Holgate, bringing the death toll from the collapse to seventy-six.

Galbraith and Kerry visited the house of Charles Davis in New Liverpool to take his evidence. Davis, married with several children, was thirty years old, and had suffered serious injuries to his back and hip; he would never work again.

All of the men described where they were working on the structure at the time of the collapse and tried their best to recount what happened. Most of the survivors were working at the leading edge of the structure, out over the water. They had all heard rumors about the bent chords. Some had seen the bends for themselves, and many had seen Birks and Yenser examining a chord. When they'd asked their foremen about it, they'd been told the same thing—the bends had always been there, and there was nothing to worry about. But the men had been concerned.

Culbert, a laborer, had taken the day off and was walking along the south shore with a fellow worker, Chase, when the bridge fell. Galbraith asked him, "Do you know of any defects in the bridge before it fell?"[5]

"Not that I personally saw," Culbert told him. "I heard speak of them, but I didn't see them. There were some defects there, I don't know if you would call them defects or not, but it would try a man's courage when they were dropping anything from the traveller to the span…you could feel it give."

"And return?"

"Oh, yes."

"The spring?"

"The spring of it; it seemed to me a little more than it ought to be."

"Did you work out there?"

"Yes."

"Did you continue work after you noticed that?"

"Everybody worked. Us fellows work as long as there is anything to stand on."

On September 19th the commissioners, accompanied by Deans, Hoare, Milliken, Edwards, McLure, and Kinloch, went to the Belair storage yard and then to the site of the wreckage. They located chord A9-L and found it shaped in the letter *S*, but largely intact, except for the latticing; most of the rivets had sheared off. The wreckage itself between the piers had slowly collapsed further, lessening the height of the steel mass. Noises and cracking sounds could be heard as the massive steel members succumbed further to the stresses caused by the collapse. There were three bodies still trapped in the steel, visible but unreachable; they could not be extracted until more of the debris was removed.[6]

An editorial in *Engineering News* appeared that day, commenting on the result of the coroner's inquest and the work that lay ahead for the Commission: "The jury concluded its work without shedding any new light on the circumstances of the accident, and the government investigating commission of engineers has begun its sessions. Naturally, a chief point of importance in these official investigations is to determine the degree of culpability of various individuals for the failure to stop work on the erection in time to save the lives of the employees. To this end, the commission has been trying to ascertain just how much evidence the bridge gave of its over-strained condition prior to the collapse."[7]

The Commission reconvened the following day, Friday the 20th. Raoul Lafrance was present and took the stand. "Did you find the plate?" Holgate asked him.

"No, there was no means of doing so," Lafrance told him. "There was too much iron on it. It is all in ruins. There is no means of finding it." He told Holgate he had spent forty-five minutes looking for the plate.

Holgate gave up. "We will dismiss this witness and trust to our own examination."[8]

Ephraim Kinloch, the Quebec Bridge Company's inspector, then took the stand. He described the circumstances surrounding his discovery of the bends in chord A9-L on Tuesday, August 27th, and the events that followed. Kinloch testified that the last time he inspected A9-L was at four o'clock on Thursday the 29th. He didn't notice anything different about the chord at that time, but he hadn't taken any measurements. He inspected the opposite chord, A9-R in the east truss, at eleven o'clock that morning, and it had appeared to be all right.[9]

Professor Kerry told Kinloch that one of the riveters, Alexander Beauvais, testified that just prior to the collapse, his partner had noticed a broken rivet at the splice between chords A9-L and A10-L, and that within minutes, another one broke. Kerry asked him, "In ordinary practice, is the failure in a relatively short time after driving of field rivets a usual thing?"

"No, it is not a usual thing, but it happens sometimes," Kinloch replied.

"In addition," Kerry said, "he said he saw the ribs bending away from the cover plate."

"The side cover plate?"

"The cover plate on one of the two inside ribs he was working on. What would you have thought of that?"

"If I had seen it, I would have been going yet," Kinloch would have fled.

Just minutes before the collapse, the rivet boss, Slim Meredith, had looked down inside that splice and told Beauvais that it wasn't any worse than the others. He didn't think it was serious.

McLure took the stand once again and provided a history of the problems experienced during the erection of the superstructure. He testified that the plate that had been referred to previously, by two witnesses at the inquest and by Lafrance at the inquiry, was in fact crimped, not cracked. The Commission moved on.[10]

Professor Kerry asked him, "To the best of your knowledge, no members went into the bridge with deformations in excess of the neighborhood of half an inch?"

"No, sir. We stretched a line, from one end of the chord to the other, between the splice or batten plates." He described the events that took place once the bend in A9-L was discovered by Kinloch, and the discussion with Yenser over moving the traveller forward.

Kerry asked him, "You and Birks were both trained engineers, Mr. McLure; did it not occur to you, in connection with that investigation, that a column that had been bent out of line under stress along its axis is likely to go on and continue to bend out of line under the same stress?"

"Yes, sir."

"At a more rapid rate?"

"In my own mind I did not consider it would do any particular harm to move the traveller."

"You did not think that the increase in the stress itself—"

"Would be sufficient to cause any trouble with that chord. To prove that, I walked out behind the traveller while they moved it."

Kerry paused, looked carefully at the young engineer, and continued. "After the traveller was moved on Wednesday morning, did you discuss whether it was necessary to take any action in regard to the safety of the bridge?"

"Nobody had any idea that the safety of the bridge was in danger."

"So that matter was not discussed?"

"Not discussed in that light."

"You are clear on that, Mr. McLure? You mean that it never occurred to anyone that the safety of the bridge was threatened?"

"No, sir."

"And that any drastic action to protect the bridge was therefore not thought about?"

"That is the idea," McLure replied in his soft voice.

"And between Tuesday morning and the time you left for New York, no accurate measurements were made to see whether that deflection was increasing?"

"No, sir, no actual measurements."

McLure told the Commission of his meeting with Cooper and that he'd made him aware that the traveller had been moved out, but that, as far as he knew, there would be no more metal erected until advices were received regarding these chords. He had asked Cooper whether, in his opinion, it would be all right to go ahead with the erection with the chords in that condition, and Cooper had sat down and wrote a telegram to the Phoenix Bridge Company.

"He did not answer you directly?"

"No, I do not think he answered me."

Kerry asked him about Yenser's decision to move the traveller out.

"I asked him his reason for doing so," McLure said, "and he said that he had too many men out."

"Practically moved the traveller out to find work…," Kerry added, shaking his head.

"For the number of men he had out," McLure continued.

"That was his reason for it?"

"Yes, sir."

"And that you advised him at the time…"

"That I thought it was poor policy."

"But at the same time, you did not consider that it was dangerously increasing the strain on the member?"

"I do not think I told him that."

"But you agreed to that with Mr. Birks."

"I had that in mind."

"You and Mr. Birks had looked into that and that was your joint conclusion?"

"Yes, sir."

"And that when you reached Mr. Cooper and gave him the facts, he immediately wired to cease increasing the load on the members?"

"Yes, sir."

Kerry asked him about the telegram from Birks that came to him while he was waiting for Cooper to arrive at his office in New York. McLure had it with him and read it: "I do not think we can state positively that chord has buckled since erection; the only evidence we have shows the contrary. See my letter with additional data in Phoenixville tomorrow morning."

"Did you see the evidence?" Kerry asked him.

"I have not seen any evidence yet that I consider positive."

"Did you see the letter?"

"Mr. Deans showed me the letter. Birks stated his reasons why he thought it was possible that the bend might have been there longer than we thought it had."

"But without anything absolute in the way of measurements."

"Yes, without anything absolute in the way of measurements."

McLure described his meeting with Deans in Phoenixville and their decision to wait for the arrival of Birks's letter the following morning before taking any action.

"Then what followed?" Kerry asked.

"I left the office about six o'clock, and I heard of the collapse of the bridge about 7:30."

"How did that word reach you?"

"It reached me by someone telling me that someone had telephoned them that the bridge had fallen down."

Twelve days of the inquiry had elapsed. It was September 21st, and for the first time, the Commission raised the issue of Cooper's telegram.

Kerry asked McLure, "Did you have any verbal instructions at all to the Phoenix Bridge Company?"

"Mr. Cooper told me to go to Phoenixville, see Mr. Deans, and tell him that some steps must be taken to strengthen that chord—I think he said—or to repair the chord, and I do not think that I told him that that evening at Phoenixville."

Kerry continued, "Mr. Cooper's telegram had already been received when you reached Phoenixville?"

"Yes, I think Mr. Deans said that he had had a telegram from Mr. Cooper when I reached there."

Stuart, the lawyer for Phoenix, spoke up. "I thought Mr. McLure said he showed Mr. Birks's telegram to Mr. Deans."

McLure answered directly. "Showed it to Mr. Cooper. I do not know whether I showed it to Mr. Deans or not. Maybe I did."

Stuart attempted to clarify. "At that time Mr. Birks's letter to Mr. Deans had not been received. It was only received on the following morning, and they decided to wait until they received the letter."

Kerry: "That is what Mr. McLure says in his evidence."

Stuart: "Much of it I miss because he speaks so low."

Stuart's interruption regarding the telegram may have been the reason why the critical question—*did Deans and McLure discuss Cooper's instruction to add no more load to the bridge?*—was not asked by Kerry. Stuart's meddling surely did not help. Kerry never raised the question, nor did anyone else.

Kerry continued, "Have you heard anything further from Mr. Cooper since your interview with him in New York?"

"I stopped to see him on my way back here."

"Did he express any opinion at all?" Kerry asked him.

"He said: 'Well, it's that chord.' I only saw him a few minutes; he was not feeling very good."

Kerry asked McLure about the system of inspection, and chord A9-L in particular. "You are absolutely convinced that you would not have passed a member with any such deflection upon it?"

"Yes. Of course, I was not here when A9-L chord was erected. Mr. Kinloch was."

Kerry turned to Kinloch who was sitting in the gallery. "Previous to Mr. McLure coming here, did that system of inspecting members on the car before they were placed in position, exist?"

"Yes, sir," Kinloch replied.

"So that chord A9-L would have been inspected by you previous to unloading from the cars to the traveller?"

"Yes, sir, and it was also free from any upright members for five or six months, and anyone walking over it could have seen it."

Kerry turned his attention back to McLure. "Had you made, for any reason, any definite observation of the chord previous to the time of the measurement on the 27th?"

"Yes, sir, of A9-L. Two or three days before I went to the hospital in early August, I sighted along each rib of the chord, particularly the rib that had the mark of the chain on it, to see if there were any bends noticeable, but particularly to see how the bend made by that chain was acting under load. And from my observation then, I am convinced that the ribs were straight, to within a deflection of half an inch."

The Commission adjourned.

Hoare was called again on the thirteenth day, Monday, September 23rd. He testified that he had been in daily communication with McLure or Kinloch by telephone. He had a private telephone in his house and there was a phone in the Phoenix office at the site. He told of his conversation with Yenser after the foreman had moved the traveller out. According to Hoare, Yenser seemed very satisfied with his decision.[11]

Davidson, the workers' lawyer, stood. "I would like to know if Mr. Yenser actually said that or whether Mr. Hoare simply thought he gave that impression."

Hoare replied, "Yes, he told me most distinctly."

Kerry asked Hoare why the report of the deflection in chord A9-L did not reach him until twelve hours after it was discovered. "It was perfectly possible, for example, Mr. Hoare, to call you up at nine o'clock that morning and let you know there was trouble."

"Yes."

"And that was not done."

"That was not done—no."

"And no effort was made to call you till after dinner in the evening to advise you of it?"

"Yes. Mr. McLure called me up and said that he was coming in to see me to show me a sketch. I do not see that calling me up earlier in the day would have done any good, because after discovering the deflection, necessarily they had to get the information to make a sketch to show all the points of deflection so as to be able to send it over by mail that day to New York and Phoenixville."

Kerry glared at the chief engineer and asked him, "Do I understand, Mr. Hoare, that if that information that Mr. Kinloch gave Mr. McLure had reached you at nine o'clock in the morning, you would not immediately have stopped everything and gone out on the bridge to inspect that yourself?"

"No, I should have required more information before I should have taken any action on it. That is the information I gave them in the evening."

"I mean personal inspection, which does not depend on the action of subordinates."

"If they had reported to me that anything serious had presented itself, I should have gone out there, but I could not have done anything without getting more particulars and that is what they were getting during the day."

Kerry continued. "Both Mr. Kinloch and Mr. McLure testified that they were seriously disturbed by this occurrence, and we understand that they took the full responsibility of not reporting that matter for the course of an entire day."

"Yes, they did, and I imagine they considered that it was not necessary to report it, as I said before, until they got complete data to lay before me as well as Mr. Cooper."

Kerry moved on; he asked Hoare about Yenser's decision to move the traveller out. "You did not take any responsibility or give any definite instructions, either to one effect or to the contrary, concerning the movement of the traveller?"

"No." Hoare replied.

None of the commissioners asked Hoare about the two letters he had sent to Cooper; the first stated clearly that Yenser asked Hoare if it would be all right to move the small traveller out and continue with the erection. In his follow-up letter to Cooper, Hoare "wished to correct a misstatement" and to clarify that he did not ask the foreman to continue the work.

Finally, Kerry asked Hoare if he had inspected chord A9-L himself after it was repaired in 1905. Hoare confirmed that he had. Kerry, "So that you knew it went in the bridge in good condition?"

"Yes," was Hoare's reply. No one asked him why he had put faith in Birks's assertion that the chord was bent when it had gone into the bridge.

The next witness was Irvin Meeser, the inspector for the Quebec Bridge Company at the shops in Phoenixville.[12]

Kerry examined him. "Previous to the fall of the bridge, Mr. Meeser, there was a discussion as to whether a certain chord member was considerably bent before it left the shop or not. What evidence have you bearing on a point of that kind?"

"I have no evidence but just what I have heard in conversation since I came over here. I have no evidence but what I found out since I came here. I found out more about it here than I did there."

"Was it your business as one of the shop inspectors to see that the chords were as perfectly straight as they could reasonably be made? Suppose a chord had not been made reasonably straight; would you have a record of that fact?"

"Yes, sir, we would, but there were none of them that ever went out, but for what were reasonably straight. We had cut chords apart before they were milled that we did not think were straight, but none of them ever passed out as finished but what we thought were reasonably straight."

"So that you were satisfied that every chord member that was shipped—"

"I am satisfied that every chord was straight. There may have been a rib that had some wave in it, but as a chord, the chord was straight."

"You tested in what way?"

"With our eye."

"You looked directly along the whole line of the chord?"

"Yes, sir."

"And you would expect to detect a wave of that amount?"

"Well, I think I could easily detect anything over half an inch, easily."

Davidson, the workers' lawyer, spoke up. "It has been in evidence that Mr. Birks was strongly of the opinion that that bend which was discovered in the chord had always been in it, that is, that it came from the shop in that condition. I would like to know if Mr. Meeser agrees in that position."

Kerry responded, "I think Mr. Meeser has already expressed himself."

"He has as a matter of fact," Davidson replied. "It is just to put the two side by side. He has already, I know, said they came away straight, but it is evident, of course, that he does not agree with the other opinion since that was his opinion."

Kerry summarized it. "It seems absolutely clear that if any crookedness existed in any one of those chords it was certainly not seen by Mr. Meeser and that he specially inspected the chords to see if anything of that sort existed."

The Commission adjourned.

By Tuesday, September 24th, public interest in the inquiry had waned. The only people present in the courtroom were the commissioners, the lawyers, the witnesses, and a few reporters. This was to be the last day of the first round of hearings in Quebec City.

Deans took the stand and explained how his engineering department had planned for the erection of the superstructure.[13] Holgate asked him about the men he had placed in charge at the bridge site. Deans replied, "I had absolute confidence in the men in charge of that work."

"And you would carry that full confidence to the extent of allowing the men there on the work to act in the case of any emergency?"

"I should expect them to act in the case of any emergency where they did not feel it was necessary to report the matter to the Phoenixville office."

"You felt that they were competent to know when an emergency arose?"

"Yes, sir, I did."

"Then, Mr. Deans, was your organization composed and carried out on the assumption that emergencies would arise?"

"We expected that they might arise during the construction of that work."

"And having that very thought in your mind, you reposed in your staff the confidence you have shown?"

"The staff was the best that we could possibly secure, and we had every confidence in them."

"If we understand the organization correctly, you even then had no man on the work who would act in an emergency or who felt competent to act in an emergency without consulting the office in Phoenixville?"

"I cannot see how we could have improved on that organization and taken care of an emergency any better except by moving the entire Phoenixville office to the Quebec Bridge. In other words, we had a force there that we thought could act in any emergency that might arise, and in which they did not have time to report to the Phoenixville office."

Deans told the Commissioners of his confidence in Birks and his last conversation with him on the morning of the collapse. Birks had convinced him that the bends in the chord were longstanding and had been there when the chord went into the structure. Deans said he had spoken to Edwards, another inspector for the Quebec Bridge Company about the chords, who told him there were "waves" in the chords and the foreman at the plant confirmed this. Deans said that, in light of the fact they were going to have a conference with Cooper the next morning, he did not think there was anything to be alarmed about. Also, they were waiting for Birks's letter to arrive on Friday.

Holgate asked Deans about events just prior to the collapse. "You say you expected Mr. McLure on the 29th. What time did he arrive in Phoenixville?"

"After our talk with Mr. Yenser and Mr. Birks over the phone on the morning of August 29th at about 10:30 a.m., I went to Philadelphia on the 11:09 a.m. train and returned to Phoenixville about three o'clock. Either then, or immediately thereafter, I received a message from Mr. Cooper's office advising me that Mr. McLure would be at our office at five o'clock. I then advised Mr. Szlapka, the designing engineer, and Mr. Milliken, superintendent of erection, to come into the office and await Mr. McLure's arrival. He reached there at five o'clock and reported his meeting with Mr. Cooper, and I asked him if Mr. Cooper had given him any further instructions, and he said no; he evidently wanted to look into the matter further. I asked him if he made any figures over there, and he said no, there was not time."

Professor Galbraith spoke up. "If he had made any figures?"

"Any calculations," said Deans.

Galbraith: "If Mr. Cooper had done so?"

"He said no," Deans told Galbraith. "There was not time; he had just told him to go to Phoenixville."

Holgate resumed. "Up to that point had there been any communication between Mr. Cooper and Phoenixville that day?"

"Just that message—the message that Mr. McLure has put in exactly—I have not a copy— advising us that Mr. McLure would be there."

"There was no telephonic communication?"

"No telephones, no other messages, and no letter."

There was no mention of Cooper's instruction to add no more load to the bridge. Not one of the commissioners ever asked Deans about it. No one did, not even Davidson, the lawyer for the workers.

Hoare was the last witness to appear for this session in Quebec. Holgate asked him, "You stated in evidence yesterday that you did not personally examine chord A9-L. Have you any explanation to account for this?"[14]

Hoare replied, "Having full confidence in Messrs. McLure and Kinloch, I depended entirely upon their investigations and measurements in all matters of that kind. To personally reach that chord, it would be a great physical effort attended to by a considerable amount of danger, unless one was in daily practice in doing that kind of work. The inspectors were there for that special purpose, and if I had to climb to look at every detail on the bridge, I might just as well have been an inspector myself and their services would not have been necessary. My work was more general in looking after the Company's business and seeing that the work was being carried out according to contract and specifications and that the inspectors on the work and at Phoenixville were fulfilling their various duties from time to time and giving me the necessary information required for the proper conduct of the work and for its monthly estimates for progress payments."

The latter, no doubt, took up most, if not all, of his time.

Holgate asked the lawyers if they had any questions for Hoare or anyone else. No one did. The Commission declared that it was concluding its hearings in Quebec, and that it would travel that night to Ottawa, then on to Phoenixville and New York.[15]

CHAPTER 21

THE INQUIRY ENDS

DAY FIFTEEN OF THE INQUIRY opened in Ottawa on September 26th. The first witness to appear was Collingwood Schreiber, the former deputy minister and chief engineer of the Federal Department of Railways and Canals; he had retired from the position in July 1905.[1] Schreiber described the steps leading up to the Order in Council giving Cooper a free hand in the design, as well as the procedures followed by his department for approval of the plans by the government. Holgate asked him, "In regards to the inspection of the work done at the bridge itself, who had you?"

"Mr. Cooper was really the man who looked after that. As I said before, the interests of the company and of the government were identical. He was supposed to visit it frequently."

"Were Mr. Cooper's personal visits frequent enough to ensure a complete inspection?"

"Well, I retired in 1905—there was scarcely anything done at that time on the superstructure."

Apparently, Schreiber was not aware that Cooper had never been to the site during the entire time that the superstructure was being erected, or after it collapsed.

On September 27th, the Commission opened its final day of examinations in Ottawa.

That morning, the Royal Mail Ship, Empress of Ireland, had arrived in Quebec City. Among its passengers were Mr. and Mrs. Rudyard Kipling, who had come to Canada on a holiday; they were to cross the country by train to the Pacific Coast. A number of speaking engagements were planned, but first, the renowned author was to receive an honorary doctorate from McGill University. The couple left Quebec by special train for Montreal shortly after the Empress arrived.[2]

In Ottawa, the bridge engineer for the Department of Railways and Canals, Robert Douglas, appeared before the Commission and was sworn.[3] He described the steps leading up to Cooper having a free hand when it came to the design, and to his department's minimal involvement in approving the plans. Kerry told Douglas, "You know there is now considerable suspicion in regard to the efficiency of some of the lower chord members?"

"Yes."

"You examined these plans from an engineering point of view and found them satisfactory?"

"I thought them satisfactory as far as the specification went. If they had been built according to the specification of 1901, the chords would have had a cover plate upon them."

"At the time that you looked over the plan, you were not at all apprehensive as to the safety of the structure?"

"No, not in the slightest, except that after the fact, or before the fact, there might be some criticism."

"But you made no criticism?"

"I was not asked; I made an examination, that was all."

"Would it be a fair statement, Mr. Douglas," Kerry asked him, "to say that for all practical purposes in connection with the actual design and construction of the bridge, Mr. Cooper could be considered as acting as engineer in charge for the department?"

"I should not say that he would be exactly—not as I understood it or understand it."

"I want to get at it, not formally, but as a matter of absolute fact. Any detail of construction that would be approved by Mr. Cooper, or any engineering question that would come up and on which Mr. Cooper would pronounce a definite opinion, would be settled in accordance with Mr. Cooper's opinion?"

THE INQUIRY ENDS

"I should say that is my understanding of Mr. Cooper's connection. I would not say with the government, but with the Quebec Bridge Company, because you could not get anything from the Quebec Bridge Company except from Mr. Cooper."

"And the department practically accepted any plans that carried Mr. Cooper's signature?"

"I do not know about the department. They were sent to me for examination; I examined them and then Mr. Schreiber approved them. He is the department. He takes responsibility of approving them."

"Mr. Schreiber told us yesterday in his evidence that it was generally understood that the interests of the government and the Quebec Bridge Company were alike, and that those interests were considered thoroughly taken care of by being entrusted to Mr. Cooper."

"That condition has arisen since I had anything to do with the bridge as an engineer—that is, since the specification—so that I know nothing about it."

"As far as you know, that was what you might call the general temper of the department?"

"As far as I know. Everything went."

The Commission adjourned their session in Ottawa to meet again at the call of the chairman.

On October 7th the Commission reassembled in Quebec for further examination of the wreckage and to study the plans and documents. One week later, they travelled to New York City. The evidence that was to be taken in the United States had to be reduced to writing and sworn to before a British consul.

From October 14th to the 19th, they met with Cooper at his residence. The sixty-eight-year-old engineer was examined by the Commission as his strength permitted.[4] Following his examination, the Commission formulated a series of questions for Cooper to address. His replies were dictated at his leisure, and the written testimony was reviewed and revised by the Commission and Cooper in further meetings, until it represented, as completely as the Commission could determine, the full testimony Cooper was able to give. Cooper's written answers were provided to the witnesses in Phoenixville before they gave their testimony. Also, the *Engineering News* magazine from New York published Cooper's testimony in full in its October 31st issue.[5]

Cooper's written answers began with a description of the events leading up to his involvement in the project, his initial retainer and report on the tenders, and his appointment to act as consulting engineer to the Quebec Bridge Company. He also referred to the Order in Council giving him full and absolute authority to make changes to the specifications, as he saw fit.

The contract between Phoenix and the Quebec Bridge and Railway Company was dated June 23, 1903, and the erection of the bridge began on July 22, 1905. Holgate asked Cooper whether enough time had been devoted to the preparation of the construction drawings. He replied, "Phoenix had the contract for the construction of this bridge several years before they commenced the preparation of the plans. I urged them at an early date to prepare their studies and plans for the accepted 1,800-foot span, stating that in an important work like this, very cautious and careful consideration would be required in each and every detail of the structure, and that this should be done before the rush of construction would come upon us. They gave this no attention, and practically made no steps towards preparing plans until they had completed their financial arrangement and executed their contract in June 1903."

The time for completion, as stated in the contract, was three years. Cooper protested and said it was impossible to do it in three years; at least four years would be required, but five would have been ideal. In his view, "The urgency and demand of the manufacturing side of this problem have outweighed and burdened the technical and thoughtful consideration of all the plans."

Regarding financial constraints, Cooper clearly understood that the structure was limited to the amount of funds estimated for the construction. The work was to be constructed by a private corporation, and the amount of money that they expected to have was a limited amount. The question to be decided was the possibility of building the best bridge within the financial strength of the company. According to Cooper, "The question of the best bridge was not brought up at all."

Holgate asked him, "What organization existed for the checking of the strain sheets and detail plans prepared by the Phoenix Bridge Company?"

"My own office organization, absolutely."

"At whose expense was this organization maintained and was it sufficient for the purpose?"

"At my own expense, and it was not sufficient for the purpose considering the other duties which were imposed upon me improperly."

THE INQUIRY ENDS

He was asked, "Was the inspection of the work of erection and the taking charge of that work properly part of the duties of the consulting engineer, and if not, whose duty was it?"

"It was not the duty of the consulting engineer," he replied. "It was the duty of the chief engineer and his organization, with the sole right to apply to the consulting engineer for advice upon any special problem."

Regarding the capabilities of the local staff at Quebec employed by the two companies, Cooper stated, "For a man to be qualified, in my opinion, to have the supreme local control of the erection of a bridge as important as that under consideration, I think he should have been a thoroughly technically educated and experienced bridge engineer. I regret to say that I do not think the chief engineer of the Quebec Bridge Company had those qualifications. In reference to the local control by the Phoenix Bridge Company, I do not think they had the quality of engineer that the circumstances demanded. In saying this I do not wish to reflect in any manner upon Mr. Birks, who sacrificed his life and who undoubtedly was a competent man in his line of experience; but I do not understand that he had the thorough training and knowledge of all the requirements of this structure necessary to fit him for the responsible position as the engineering representative of the contractor on such an important structure."

According to Cooper, "In the case of the Quebec Bridge Company, like all projects undertaken by men not specially acquainted with the necessities, the engineering features of any such great work, they were unable to make a proper selection. In reference to Phoenix, I think it was due to the fact that the commercial branch of the company gave more consideration to the pushing of the work than they did to the giving of due consideration to the practical requirements of such a great structure."

In Cooper's view, McLure "was the only person who had any preparation or qualifications for supervising the construction of that bridge, and I know that the time allowed him for preparation for this important duty was not as great as should have been given him. From the reports that he has from time to time sent me, from personal contact with him, I felt that he did all that could be expected of him under the circumstances."

"How often did you visit the bridge site during the erection of the superstructure?" Holgate asked him.

"Never," Cooper replied. "I have never been able to visit the bridge since the erection commenced. I was disabled before that was undertaken."

"Was it the practice of the Quebec Bridge Company's staff to refer all difficulties to you, and, if so, what were the duties of the chief engineer?"

"As far as I know, all difficulties, all questions, all decisions on any matter relating to the structure were referred to me, and practically, as I now see it, I was acting not only as the consulting engineer but as the chief engineer of the Quebec Bridge."

Cooper then told the Commission about his remuneration, and he referred to a conversation he'd had with Hoare early on. "I suggested to Mr. Hoare that it was hardly fair, considering that I had reduced my fee to one-half, that I should not be granted some additional remuneration to aid me in carrying out the duties that had been placed upon me. No such additional remuneration has ever been granted me, and no offer has ever been made to restore my original fee. My staff and office expenses due to the work required in the interest of the Quebec Bridge Company have been paid entirely from my own fee, and they have amounted to approximately the sum that I have received from the Quebec Bridge Company to cover my employment."

Asked whether he considered that there was clear indication that the failure was imminent, and whether it was possible by prompt and intelligent action to have prevented the failure, Cooper replied, "I think the deflection of an important member, as chord 9 west, to the extent of 2-1/4 inches, would indicate to any intelligent mind that the chord was less capable of doing the duty that it would have done if in a perfectly straight condition. And I do think that it was perfectly possible by prompt and intelligent action to have stayed that chord and prevented the failure of the bridge."

Holgate asked him, "By whom should the orders for such action been given, and to whose lack of judgment and initiative can the failure therefore be charged?"

"To the executive officers of either company who were present or within sufficient touch to have given any orders."

Asked whether, in his opinion, it was good practice to leave the ordering of such action to any employee of a contracting company, Cooper replied, "The contracting company should have had on the structure an employee of sufficient intelligence to have appreciated the necessity for and to have given such an order. At the same time, the responsible executive of the Quebec Bridge Company should not have hesitated, in the absence of proper action by the contractor, to have given such an order."

"Do you think that, at moderate expense, the ribs could have been made absolutely safe?"

THE INQUIRY ENDS

"I do. I believe if prompt action had been taken to protect chord 9 west from further deflection, which could have been done by the employment of three hours' work and a hundred dollars' worth of timber and bolts, the defects and deficiencies which we now recognize in the compression chords and members could, at a later date, have been corrected and the bridge could have been made safe and efficient for its intended purpose."

Finally, he was asked whether he thought the engineering data at their disposal was sufficient to enable engineers to design members similar to those in the lower chord with safety and economy, and whether he would recommend material changes in the detailing of these or any other members.

Cooper replied, "My responsibilities, gentlemen, end as soon as I have served my duty of aiding you in reaching the truth in regard to the destruction of this bridge. While I have my views, and such views are at the service of those who have heretofore relied on me, I shall decline to take any executive or responsible position in connection with the correction of the errors that we now recognize in this work. It must be referred to younger and abler men."

From October 23rd to November 22nd, the Commission took evidence and collected information in Phoenixville and Philadelphia. Disturbed by reports of Cooper's testimony criticizing Phoenix, an unnamed Phoenix official told a reporter, "Within a week Mr. Cooper's allegations will be proven false."[6]

The first witness examined was David Reeves, president of the Phoenix Bridge Company.[7] When asked about Cooper's involvement with the project, Reeves told the commissioners, "I believed the appointment by the Quebec Bridge Company of Theodore Cooper as consulting engineer assured the success of the undertaking, that our engineers and constructors were fully competent to design, construct, and erect the bridge under Mr. Cooper's supervision, and that of the engineers of the Quebec Bridge Company and the Department of Railways and Canals."

When asked generally about the events that led to the failure, Reeves testified, "The cause of the failure cannot be found due to any departure from the specifications in design, material, or workmanship, or lack of good judgment in the field. No engineer under the circumstances will accept the idea of a local defect to account for it. The profession is bound to look beyond that—in the employment of the unusually high stresses prescribed for compression members, beyond all precedent, and as it now appears, beyond the existing technical knowledge of their effect."

Reeves pointed out that Cooper stated that the Order in Council gave him absolute authority to amend the specifications, and to order such alterations in the construction plans as seemed best in his judgment. "This expression of his absolute and final authority coincided with our understanding of it in our dealings with him under the contract. He made modifications in the unit stresses to be employed upon the various members, which very much increased them beyond any precedent, and by doing so placed the whole design in a field outside the benefit of experience. Such high stresses had never before been used, and in using them he acted with the authority of the Quebec Bridge and Railway Company and the Dominion of Canada vested in him. The fall of the bridge is to be laid directly to the change in the unit stresses as made by Mr. Cooper."

John Sterling Deans, the chief engineer for Phoenix, was next. He had already testified on several occasions in Quebec City, and would be questioned several more times in Phoenixville and Philadelphia.[8]

Deans was asked, "Did the changes in the unit stresses meet with your approval?"

"The changes in unit stresses were made by Mr. Cooper and were not submitted to us for our approval. Mr. Cooper merely talked the matter over with Mr. Szlapka as a brother engineer, but not, however, for the purpose of getting the wishes of the Phoenix Bridge Company. He then reached a decision of which we were notified and upon which we acted."

"Did these changes follow previous experience, or did they take the work out of the field of past experience in bridge construction and detailing, and in what respect?"

"The changes in unit stresses for compression members carried them out of the field of past experience and did not follow usual practice."

"Was the design of details of the lower chord particularly discussed with Mr. Cooper, and was his opinion specifically obtained on the latticing and other details, and if so, please state fully what took place."

"Yes. I had no interview with Mr. Cooper on this subject, but I instructed our designing engineer particularly to submit the question of size of latticing of chords to Mr. Cooper. Mr. Szlapka later reported to me that he had an interview with Mr. Cooper on this point, and that Mr. Cooper advised him that the lattice angles were correct as shown on approved plans."

"Were chords A9-L, anchor arm, and C9-R and C8-R, cantilever arm, in perfect condition when they left Phoenixville?"

THE INQUIRY ENDS

"Yes," Deans replied. This testimony contradicted what he had said and written prior to the collapse, but he was not challenged by the commissioners.

"Please explain the references in Mr. Birks's letter of August 29th, with regard to the telephone conversation about stopping the work of erection?"

"On August 29th, we first learned from the letter of August 27th from Mr. Yenser that buckles were noticed in the webs of lower chord 9-L of the anchor arm. Consultation then took place at Phoenixville between the engineers, shop officials, and inspectors, and it was determined that the chord could not be bending from any excessive stress, as it was carrying only three-quarters of the workload for which it was designed and approved." Deans then described the telephone discussions that took place between the site and the office.

At no time did the commissioners ask Deans why he had ignored Cooper's instruction to add no more load to the bridge, but he addressed the issue himself in his response to the Commission's question about stopping the work of erection: "While a chord with bent webs, even though the bends are slight, is not capable of performing its functions as well as a perfectly straight member, the bends in chord A9-L noted on August 27,th and of which we learned on August 29th, were not such as to shake the absolute confidence of years, which all had in the entire structure. If the consulting engineer then believed there was imminent danger and that all work should be stopped immediately, it was not necessary to inquire whether Milliken was at site or not. Hoare had sent McLure to Cooper to report on the bends in chord A9-L and to receive his advice. Hoare was in Quebec and any message to him would have stopped the work instantly, as was done on a previous occasion by direction of Cooper. The testimony of others shows that Cooper, on August 29th, no doubt, had no thought of imminent danger. We all now see what no one dreamed of before, that the compression chords were beyond any scheme of protection on August 29th and were failing under less than half the load for which they were designed and approved and were considered capable of sustaining without failure. While it is difficult, it is essential, in order to reach an accurate judgment, to keep in view the frame of mind everyone was in before August 29th regarding this structure and its strength, and the respect and confidence all had in the engineers responsible for its design and detail."

The chief design engineer for Phoenix, Peter Szlapka, gave his testimony in Phoenixville.[9] After going through his background and the organization in his engineering office, he was asked, "Was the study of the design what you would call complete, having regard to its unprecedented dimensions and also having regard to the fact that details had not been fully considered?"

"A continuous study was given to the general design, while the details were perfected as the work progressed. The final design, I believe, cannot be improved upon."

Szlapka stated that when the detail design was commenced in 1903, "The weights of the cantilever arm and suspended span were then believed to be sufficiently accurate—and were so approved by Mr. Cooper—to enable me to correctly design the anchor arm. Subsequently, when the suspended span and cantilever arm were developed, it was found that the actual weights were somewhat in excess of those assumed for the calculation of the anchor arm."

In fact, the suspended span and cantilever arm were twenty-percent heavier, and the anchor arm weighed thirty-six percent more than what Szlapka had assumed.

He was asked, "Why was not the whole scheme of the bridge fully considered in detail before the shop work commenced?"

Szlapka replied that it was not the practice of the Phoenix Bridge Company office to check the correctness of the assumed dead load. "This was not practically possible. General experience enabled us to proceed without occupying valuable time, and the time limit precluded any such arrangement. This followed the usual course of business in such cases."

He was then asked, "Having in view the unprecedented dimensions of this structure, was it the proper course to pursue, or did you pursue the ordinary course as followed previously in connection with bridge building?"

"The ordinary rule, which is imperative in all cases, irrespective of the unprecedented dimensions of this structure, was followed."

"Whose instructions did you follow in adopting the above course, and what were the instructions?"

"I received my instructions from Mr. William H. Reeves, general superintendent, and Mr. John Sterling Deans, chief engineer of the Phoenix Bridge Company; that is, to place with the shops any shop plans as soon as approved, and to generally arrange the office work so as to ensure continuous working on the bridge, in the shops, and in the field."

Regarding the unit stresses specified by Cooper, Szlapka told the Commission, "In view of Mr. Cooper's proposition to use, for certain combinations of conditions, unit stresses as high as 24,000 pounds, or three-quarters of an average elastic

THE INQUIRY ENDS

limit of 32,000 pounds, I mentioned to him the fact that a German professor proposed to use a fraction of the elastic limit for unit stresses for truss members after first allowing for irregularity of shop work, for imperfect erection, for flaws in materials, etc."

Cooper, apparently, had not agreed.

"Did you fully concur in all amendments made in the specifications, having in mind that you were endeavoring to produce the best possible bridge?"

"The amendments made in the specifications by Mr. Cooper were not subject to my approval," Szlapka told the Commission.

Szlapka visited the bridge in 1901, 1905, 1906, and in August of 1907. In his view, the weakest part of the structure during erection, and when completely erected, were the compression members, yet he never examined any of the chords on any of his visits, including his last visit, when he was aware that the chord splices were proving to be difficult. Of course, it would have taken some effort to climb down fifty to seventy-five feet into the structure to get a firsthand look at these critically important members.

He told the commissioners, "It was impossible for me to believe that the bridge was failing or that the amount of curvature in chord A9-L was as reported. Our resident engineer, Mr. Birks, stating on August 29th, on the telephone, that there was no distortion in any lattice, that all rivets were tight, that there was no change taking place in any part of the chord. I was further strengthened in my belief that there was nothing wrong with that member. I made rough calculations of the chord, however, using 14,000 pounds axial stress, and an average curvature for the four ribs of the chord of 1-1/2 inches, and found that even with this improbable curvature, the chord was not in a dangerous condition."

In fact, the ribs were bent 2-1/4 inches.

Szlapka described his state of mind on August 29[th.] "Knowing that every part of the bridge was figured with the utmost care as to its strength, that the results of the calculations were checked and compared at least three times in the Phoenix Bridge Company's office, that they were then sent to the consulting engineer for comparison with his calculations and for his approval, and that they were fully approved by him; knowing further that the shop plans were prepared under my personal supervision by a corps of able engineers and draftsmen, that these plans were redrawn several times, that they were then sent to the consulting engineer for his study and approval, and that they were all approved by him; knowing further that every part of the bridge was constructed strictly in accordance with these plans; knowing also that the erection was conducted carefully and strictly according to plans prepared by the engineering department—knowing all these facts, I was forced to believe that on August 29, 1907, the bridge was in a safe condition, and that no part could show the least sign of weakness due to stress, especially as the loads of the bridge on that day were such as to produce stresses in the truss members only about three-fourths of the stresses the bridge was figured to be able to bear, with entire safety, after its final completion."

The Commission, having for the time being concluded the inquiry in New York, Philadelphia, and Phoenixville, returned to Montreal and spent time examining and discussing the evidence, and preparing their report. A second visit was paid to Quebec on November 28th, for the purpose of re-examining Hoare and Kinloch, and pursuing other matters.[10]

During his re-examination, Hoare was asked, "Why did you not stop the work on the bridge on August 28th, pending Mr. Cooper's decision, and with the information you had in regard to the condition of some of the compression members?"

"I did not stop the work on August 28th for the following reasons: I did not consider the conditions warranted such action, particularly as the Quebec Bridge Company's inspectors and the Phoenix Bridge Company's engineer and foremen disagreed upon the origin of the deflection. The latter showed no signs of uneasiness and were anxious to continue the work, as they had made a special effort to collect a large force of bridge men. As I understood it, the majority of men were engaged removing the large traveller and riveting and they would add little extra load until expected instructions were received from the consulting engineer upon Mr. McLure's arrival. My confidence was strengthened by the knowledge that very careful work had been performed by expert designers who had been entrusted with the calculations and preparations of the plans of the bridge, and that at the time the chord was not strained over three-quarters of the maximum provided for and that a mistake was impossible under such conditions; and it was also reported to me that the ribs had a full bearing at the splices."

Hoare had previously testified that he had personally seen chord A9-L after it had been repaired, in the company of several others, including Szlapka, and was certain that it was straight when it went into the bridge. None of the commissioners reminded him of this.

THE INQUIRY ENDS

Kinloch was re-examined regarding the events that occurred following his discovery of the bent chord on August 27th.[11] He told the Commission that when Hoare came to the site on the 28th, "He appeared very anxious that I should abandon my position of being positively convinced that the bend in chord A9-L had occurred since the erection of the cantilever arm was completed, and argued both this and some possible methods of strengthening the chords by bracing several times with me. I was somewhat excited and much annoyed at the unwillingness of all the engineers to accept my statement of facts and on both Wednesday and Thursday avoided further discussion of the matter as much as possible. It was understood that McLure would immediately wire me if Mr. Cooper took a serious view of the situation, but this he failed to do."

At no time during any of McLure's examinations did the Commission raise his alleged promise to wire Kinloch.

On December 3rd Holgate travelled to New York to further examine Cooper.[12] He asked him, "Do you assume full responsibility for the change in the specifications, and for the selected unit stresses?"

"Yes," Cooper replied.

"Will you please say why, when you telegraphed the Phoenix Bridge Company at noon on August 29th, you did not telegraph also to the chief engineer of the Quebec Bridge Company? We understand that on a previous occasion you stopped work on the bridge by adopting this course?"

"During the half hour that I had this matter under consideration I felt that prompt action was needed to stop any more loading and to promptly protect the chord from further deflection. Learning from Mr. McLure that there was no one upon the work but the foreman, realizing that it might be very slow reaching Mr. Hoare, as he might be at his home, his office, the bridge, or some other place, I decided that the shortest and quickest method of reaching the bridge was through the Phoenix Bridge Company, who, I knew, had direct wire and telephonic communication with their office at the bridge. On the previous occasion when I stopped work on the bridge by communicating with the chief engineer of the Quebec Bridge Company there was no emergency before me."

Finally, on January 4th, 1908, Hoare and Kinloch were examined one last time in Quebec on some technical matters, thus ending the examination of witnesses.[13]

Holgate had sent a letter to Simon Napoleon Parent, the president of the Quebec Bridge Company, asking him to state what he considered to be the real duties of Hoare and Cooper, and what he as president of the Company expected from each. Parent wrote, "While Hoare had the title of chief engineer and Cooper that merely of consulting engineer, still we considered the latter as being in fact chief engineer of the enterprise. At the time the services of Mr. Cooper were secured, he would not undertake this work unless given full control over it, not only in the preparation of the plans, but also during the execution of the work."

Parent gave several examples where Cooper exercised his authority over the work, including his instruction to Hoare in 1906 to stop the work, and of Hoare's inability to make a decision two days prior to the collapse, when he sent McLure to New York to get advice from Cooper. Finally, he noted that on the day of the collapse, Cooper sent his telegram to Phoenixville, not to Hoare, "as would have been the proper course if the latter had been the one in authority."[14]

The failure mechanism of the collapse was described by Cooper, Szlapka, and Deans, but, in the opinion of the commissioners, it was Kinloch who gave the best description.[15] "The initial failure, I think, occurred in both lower chords No. 9 of the anchor arm simultaneously, and in the latticed portion of the chords, but not in the same way in both chords. The four ribs of chord A9-L, which had previously been observed to be bent, deflected slowly and transferred some of its load to A9-R, until that chord burst with a sudden fracture accompanied by a the loud report testified to by some witnesses. The sudden and complete collapse of A9-R whilst A9-L was slowly yielding accounts for the slight swing of the cantilever arm downstream towards Quebec, and for the tendency of the upper portions of the anchor arm to fall in the same direction. At the moment of the collapse the thrust of the cantilever arm forced the feet of the main posts off the pedestals and the shoes of the main posts were the first part of the structure to strike the ground..."

It was clear from the evidence that the top of the tower fell slowly outward, towards the river, following the cantilever arm and the large traveller into the St. Lawrence River. The cantilever arm and suspended span fell almost horizontally, in one piece.

From the evidence of the timekeeper, Huot, the Commission was able to estimate the duration of the fall at fifteen seconds. Huot ran from the second panel of the anchor arm to the office at his topmost speed, a distance of almost one hundred yards. The floor was already opening up between the end of the anchor arm and the approach span as he passed that point.

THE INQUIRY ENDS

While the commissioners were in Phoenixville in November, the Phoenix Bridge Company, on its own initiative, conducted a series of compression tests on a one-third-scale model of the lower chord 9 of the anchor arm. The model carefully replicated the details of the full-sized member, and in particular, the latticing. When the loading on the chord section reached 26,850 pounds per square inch, it failed with explosive violence by the shearing of the latticing rivets, followed instantly by the buckling of the four ribs of the chord. Further tests were conducted in late January 1908, on another chord section with latticing that was twice the strength of that of the previous test. The chord withstood a load of 37,000 pounds per square inch before it failed.[16]

CHAPTER 22

THE FINDINGS OF THE COMMISSION

THE COMMISSIONERS COMPLETED THE INQUIRY in early January 1908. A total of forty-three witnesses had been examined, many of them several times. A.E. Hoare appeared before the Commission eleven times. The report itself was only five pages long, but it attached nineteen appendices addressing each subject matter investigated by the commissioners. The entire document totaled nearly a thousand pages, in four volumes. The transcript of the testimony was over four hundred pages long.

The report was presented to Matthew Joseph Butler, deputy minister of the Department of Railways and Canals, on February 20th, 1908, and tabled in parliament on March 9th. It began with a description of the Commission's activities since its appointment, and set out what they considered to be their mandate: "To determine to the best of our ability the cause of the collapse of the Quebec Bridge, and to thoroughly investigate any matters appertaining thereto which might enable us to explain that cause."

Five days after the report was sent to Butler, Chairman Holgate sent him a letter informing him that his findings were "not completely in accord with those of my colleagues." Holgate had intended to present not merely "an analytical description of why the bridge fell, but what might be called a semi-technical and semi-business description of the whole of the occurrence." His two colleagues, on the other hand, wanted to focus on "purely technical matters." The conflict centered on the appendices, which Holgate told Butler were meant to "explain the findings." He wrote, "My colleagues now are apt to define the terms of the Commission in what I consider to be a narrow way, and do not desire some of the information which I have, and which I think should be stated, to be transmitted through these appendices."[1]

Appendix 1 was the transcript of the evidence given before the Commission, appendix 2 comprised the exhibits and appendix 12 gave a description of the fallen structure. Holgate had prepared appendices 3 through 11 and sent them to his colleagues in December 1907. Holgate's appendices addressed the history of the companies involved, the development of the specifications and plans, the procedures followed in the Phoenix office, a description of the organizations involved and their staff, as well as the effect of financial limitations on the design of the bridge. They also discussed the difficulties that arose during erection and, finally, the events that took place just prior to the collapse.

Galbraith and Kerry disagreed with Holgate as to the form in which this information should be presented, and presumably, Holgate's interpretation of the facts. They had prepared appendices 13 through 19, which addressed technical matters only. These included an examination of the various column tests performed, a comparison of the stresses as designed with the stresses authorized by the specifications, a discussion of the theory of built-up compression members, and a comparison of the design for the chords of the Quebec Bridge with those of other cantilever bridges. Holgate wrote that he had, "from the start, opposed the preparation of them (appendices 13 through 19) and also the occupation of time in gathering the information necessary for them, for the reason that I do not think we had anything to do with the building of a new bridge, and I cannot see but that these have been considered from that point of view."

In the end, all nineteen appendices were attached to the report. Holgate, however, considered the technical appendices to be subordinate to his. The findings of the Commission suggest that Holgate was less than successful in emphasizing the "semi-business" side of things.

The commissioners made fifteen findings. They are best understood when read together with the relevant appendices:[2]

> (a) *The collapse of the Quebec Bridge resulted from the failure of the lower compression chords in the anchor arm near the main pier. The failure of these chords was due to their defective design.*
>
> (b) *The stresses that caused the failure were not due to abnormal weather conditions or accident, but were such as might be expected in the regular course of erection.*

Appendix 11 of the report, as drafted by Holgate, states, "As a conclusion reached from the evidence and from our own studies and tests, we are satisfied that the bridge fell because the latticing of the lower chords near the main pier was too weak to carry the stresses to which it was subjected, but we also believe that the amount of those lattice stresses is determined by the deviation of the lines of centre of pressure, from the axis of the chords, and this deviation is largely affected by the conditions at the ends of the chords. We must therefore conclude that although the lower chords A9-L and A9-R of the anchor arm, which, in our judgment, were the first to fall, failed from weakness of latticing, the stresses that caused the failure were to some extent due to the weak end details of the chords, and to the looseness, or absence, of the splice plates, arising partly from the necessities of the method of erection adopted, and partly from a failure to appreciate the delicacy of the joints and the care with which they should be handled and watched during erection. The Phoenix Bridge Company showed indifferent engineering ability in the design of the joints and did not recognize the great care with which these should be treated in the field."[3]

The commissioners attributed the failure to "indifferent engineering ability" on the part of the Phoenix designers, and to a lack of care in the field. Neither the findings, nor Appendix 11, adequately addresses the influence or impact that the organizational structure set up by the project participants had on the design and erection of the structure.

Holgate appeared before a parliamentary committee looking into the financial aspects of the bridge failure in July 1908. He was asked how such a bridge could be designed to ensure success. Holgate replied that he had given the matter a good deal of consideration, and he told the committee, "The new work must be controlled by a more competent and conservative organization." He recommended that the design and construction be entrusted to a commission of three of the most eminent bridge engineers in the world, independent from the contractor, and paid by the owner. Although Holgate's suggestion for the second bridge was followed, his opinion was not reflected in his own report.[4]

In her thoughtful and thought-provoking paper, titled "Fixing the Blame: Organizational Culture and the Quebec Bridge Collapse," Professor Eda Kranakis from the Department of History at the University of Ottawa, argues that the conventional view of the collapse is flawed, and that "the errors behind the collapse were rooted in the project's organizational culture."[5] Her arguments are unassailable. During the early stages of the project and throughout its design and construction, the three organizations involved in the project—the federal government, the Quebec Bridge Company, and Phoenix Bridge Company—made, or failed to make, decisions that set in motion a series of events that led directly to the collapse.

> *(c) The design of the chords that failed was made by Mr. P.L. Szlapka, the designing engineer of the Phoenix Bridge Company.*
>
> *(d) This design was examined and officially approved by Mr. Theodore Cooper, consulting engineer of the Quebec Bridge and Railway Company.*
>
> *(e) The failure cannot be attributed directly to any cause other than errors in judgment on the part of these two engineers.*
>
> *(f) These errors of judgment cannot be attributed either to lack of common professional knowledge, to neglect of duty, or to a desire to economize. The ability of the two engineers was tried in one of the most difficult professional problems of the day and proved to be insufficient to the task.*

The use by the Commission of the term "errors of judgment" is noteworthy, particularly in light of the wording in finding (f) that these errors of judgment did not arise from negligence or lack of knowledge. In the Commission's view, these were honest mistakes for which the engineers could not be held liable—they did their best, but the task was simply too great for them.

An error of judgment may or may not constitute negligence; it depends on whether the person who made the error was exercising reasonable care at the time the error was made. At the time the chords were designed by Szlapka, there was no established system in place for designing large compression members; the individual judgment of the engineer was the determining factor. There was also practically nothing in the specifications to guide Szlapka; he based his design on chords for much smaller structures. The only aid he had was Cooper's own specification that stated that the lattices tying the ribs together had to be proportional to the size of the ribs. The only way to determine whether the design was adequate was to test the members as designed.

Cooper approved the design and subsequently checked it again after being told by Szlapka that Reeves, the president of Phoenix, had expressed the opinion that the lattices appeared light. When Cooper checked the design again, he told Szlapka that they "had it all right." Cooper didn't order any tests of the compression chords as there was no testing machine large enough to make full-size tests. Following the collapse, however, tests were conducted on one-third scaled down versions of the chords, on Phoenix's own testing machine, the largest in the world. The test results replicated almost exactly the failure of chord A9-L.

In Appendix 16, the Commission observed, "The main criticism that can be made of the designers was that they had the means of checking their theories by use of the testing machine and that they did not do this, nor did they thoroughly study the possibilities of lattice formulas." Cooper testified that he failed to give the lower chords the degree of personal attention that he gave to the details of the tension system. Hundreds of tests were conducted on the eyebars that comprised the upper chords of the trusses. These had been designed by Cooper himself, following several failed attempts by the Phoenix engineers to satisfy the consulting engineer. There were over 700 eyebars in the wreck, and only one was found to be broken. Both engineers testified that they considered the lower compression chords to be the weakest part of the structure, yet they paid little or no attention to them.

The circumstances were such that this was to be the longest span bridge ever built. While every effort was made to design and test the upper tension chords, which proved adequate to the task, no similar effort was made to properly design and test the lower compression chords, which both engineers admitted were the weakest part of the structure. It can be argued that these errors of judgment on the part of Szlapka and Cooper resulted from a failure to take reasonable care in the design of the compression chords.

The commissioners make no mention in their findings of the desire on the part of the Quebec Bridge Company and the federal government to have the bridge completed in time for the tercentenary in 1908. There is no doubt that the careful study of the lower compression chords and their testing would have delayed the work, jeopardizing the completion date.

Regarding the Commission's finding that the errors of judgment behind the collapse did not result from a desire to economize, Professor Kranakis points out "this was true only in the narrow sense that there was no pressure on Szlapka after 1903 to design the bridge within a predetermined cost. The commission justified its interpretation only in reference to the project's post-1903 financial circumstances, ignoring the fact that the specifications and crucial organizational and technical decisions were largely finalized by that time. By 1903, the culture of thrift had already sent the undertaking down a particular technical path, which diverted all subsequent decisions."[6]

When Cooper recommended that the span be increased to 1,800 feet, he intended to add as little steel as possible to the structure. The board's approval of the increased span was conditional upon the efficiency of the structure being maintained. In other words, there could be no significant increase in cost. The weight of the steel structure was increased by less than twenty-four percent.

Regardless of whether Szlapka and Cooper should have been excused for the defective design of the chords, their unwillingness to question their design once the chords began to show signs of failure points to an air of overconfidence that bordered on arrogance. In early August, when the lower chords of the cantilever arm began to buckle, Cooper and Szlapka could not understand how the bends could possibly have occurred. It was, to Cooper, "a mystery," since the members were not yet fully loaded. He wrote to McLure on August 15th that he thought the chords had been hit by the suspended beams used during their erection and ordered him to find evidence of it. Szlapka was physically at site two weeks before the failure, but he never took the time or made the effort to examine the bent chords of the cantilever arm. In fact, Szlapka testified that he never looked at a single compression chord in the erected structure on any of his three visits to site.[7]

On August 29th, Cooper finally saw and understood the situation for what it was. The bridge was failing under its own weight and the design of the chords was flawed. The question then became whether or not he believed there was imminent danger.

> *(g) We do not consider that the specifications for the work were satisfactory or sufficient, the unit stresses in particular being higher than any established by past practice.*

In August 1904, Cooper wrote to Szlapka, after receiving the design for the chords, "I find that the only members exceeding 24,000 pounds per square inch in compression are the lower chords which have 26,500 pounds per square inch, and that is all right." He was fully aware that the lower chords would be subjected to stresses higher than the 24,000 pounds per square inch specified, far beyond normal practice. In fact, they would be unprecedented, yet Cooper paid little attention to their design, as did Szlapka.

Szlapka confirmed in his testimony that once the structure was completed, the combination of wind, snow, and dead and live loading would produce unit stresses in the lower compression chords that would exceed those permitted by the specifications. Chord A9-L was expected to carry a load of 26,800 pounds per square inch, according to Szlapka—eleven percent higher than specified. Due to its poor design and erection, it failed under a load of 18,000 pounds per square inch. The riveted lattices were simply too weak to allow the four ribs to act as one unit—they never would have developed their full strength.

The report does not address the absence of an independent peer review and its potential benefits, nor the circumstances surrounding the decision to give Cooper supreme authority. From the moment that the federal government acceded to Cooper's demand that there be no peer review, the design for the longest span bridge in the world was left to the discretion of one man. Of course, this was precisely what the Phoenix and Quebec Bridge Companies had asked for.

> *(h) A grave error was made in assuming the dead load for the calculations at too low a value and not afterwards revising this assumption. This error was of sufficient magnitude to have required the condemnation of the bridge, even if the details of the lower chords*

had been of sufficient strength, because if the bridge had been completed as designed, the actual stresses would have been considerably greater than those permitted by the specifications. This erroneous assumption was made by Mr. Szlapka and accepted by Mr. Cooper and tended to hasten the disaster.

Appendix 14 of the report states, "On minor bridges, with a given live loading, the weight of metal is known not to vary greatly with details of design, and in some offices, revision of the assumed deal load for such a bridge is not the rule; but no information from which to predict the weight of the Quebec Bridge existed, and the probability of a serious mistake in the first estimate for weight would be apparent to a cautious designer. The fact is that Messrs. Deans, Szlapka, and Cooper permitted the shops and rolling mills to commence work without taking any steps to test the correctness of the assumed dead load, and the probable dead load does not appear to have been estimated from the plans until at least eighteen months after the work of fabrication was commenced.

"We are of the opinion that no manufacturing should have been done until the designers had so far advanced with their work as to be able to make a proper estimate of the weight of the bridge. Before completing the drawings for use in the shop, the weight of the various parts should have been computed as a check on the estimated weight of the bridge. As a matter of fact, this procedure was not adopted, and manufacturing was commenced in July 1904 without any such checking, although the specifications called for it, and the contract practically demanded it.

"The designing office had accumulated sufficient information to enable it to make a close estimate of the weight of the bridge but did not do so. On the contrary, work continued as if their assumptions had been correct."[8]

Szlapka testified that it was the practice in the Phoenix office not to check these weight assumptions, and that he was under standing orders to "ensure continuation of the work in the shops and in the field." Characterizing the underestimated weight as an "error" on the part of Szlapka fails to recognize this fact—Szlapka was simply following company policy. Phoenix took a risk that its weight assumptions would be correct, all with a view to maintaining production. They had no idea what they were dealing with; the Quebec Bridge was beyond anything they had ever undertaken, yet they treated it like all of the other smaller bridges they had built, with Cooper's approval. The commissioners' findings make no mention of this, nor did they admonish the Phoenix Bridge Company for its practices.[9]

Following the collapse, the Phoenix Bridge Company made some changes to its practices. Professor Kranakis explains that Phoenix "reviewed and changed its work-flow policy. Although it claimed publicly to have been vindicated, it knew it had to rethink its design and production methods. Following the Royal Commission's investigation, it retained an engineer with the New York City Department of Bridges, Alexander Johnson, to estimate the time needed to fully prepare all detailed working plans for a bridge like the Quebec Bridge, in advance of construction. The consultant advised that at least two full years were needed to prepare general plans, stress sheets, and detail drawings, plus at least another six months to prepare the first batch of shop drawings to initiate manufacture and erection. This meant an overall delay of at least three years between the beginning of design work and the beginning of erection (in contrast, the time lag for the Quebec Bridge, on account of segmenting the design process, had been only a year and a half). In essence, Johnson's report gave the Phoenix Bridge Company the minimum time required to achieve a correct weight estimate for a large bridge. Phoenix proposed to follow these guidelines in a new contract to redesign and rebuild the Quebec Bridge."[10]

In its findings, the Commission failed to highlight the fact that the Quebec Bridge Company accepted the Phoenix production approach and that it urged the federal government to do the same. The government also acquiesced to the demand that Cooper be given supreme authority to amend the specifications. All three organizations should have been held accountable for placing unrealistic time constraints on the designers, for allowing the bridge to be designed outside the realm of standard practice, and for giving the consulting engineer supreme authority. Singling out Szlapka and Cooper for the weight issue diverts attention away from the underlying causes that led to the problem.[11]

Once the actual weight became known, however, Cooper and Szlapka refused to accept the consequences and chose instead to "submit to it," as if, by Cooper's sheer will, the bridge could withstand the stresses resulting from the increased weight. No mention is made of this by the commissioners, nor is there a discussion of the circumstances surrounding the decision by all three organizations to give Cooper the final say, without the benefit of a peer review. No doubt, an independent engineer would have questioned the decision to continue with the work. At the very least, everyone involved would have been made aware of the underestimated weight.

THE FINDINGS OF THE COMMISSION

This weight error by Phoenix also meant that they originally underbid the job. The commissioners noted, "The evidence shows that Phoenix gave more time and attention to the competition than any other tenderer, but the error afterwards made by it in assuming the weight of the structure for final designs shows how faulty the estimate accompanying its original tender was." Had the final contract between Phoenix and the Quebec Bridge Company been based on a fixed price, as originally intended, Phoenix would most likely have lost money since the final weight of the structure was thirty percent higher than their estimate, and the price per pound of steel had escalated considerably.[12]

J.A.L. Waddell was a Canadian engineer and a McGill University graduate. He had approached the promoters of the Quebec Bridge during the mid-1890s and offered to design and inspect the bridge for a total fee of $100,000. The secretary of the board had written to him, declining his offer. Waddell's approach to designing bridges differed significantly from the approach adopted by Cooper and the Quebec Bridge Company. The latter allowed the contractor to prepare the design, based on approved specifications, with the cost being included in the contractor's price for construction. Waddell, on the other hand, would have prepared the design and then called for tenders on the approved design. Of course, this would have taken more time and the cost would have been paid by the owner. Following the collapse, Waddell criticized everyone involved with the undertaking and attributed the disaster to the approach adopted. It was, in Waddell's view, the contractor's desire to put production ahead of thoughtful, detailed design that led to the collapse.[13]

> (i) We do not believe that the fall of the bridge could have been prevented by any action taken after August 27, 1907. Any effort to brace or take down the structure would have been impracticable owing to the manifest risk of human life involved.

Cooper testified that the ribs could have been made safe, "by the employment of three hours of work and $100 worth of timber and bolts." The commissioners disagreed. "Our tests have satisfied us that no temporary bracing such as that proposed by Cooper could have long arrested the disaster; struts might have kept the chords from bending, but failure from buckling and rivet shear would have soon occurred."[14]

It would have been impossible to straighten these massive chord members while they were carrying these enormous loads, or to render them secure against further deflection and collapse. This would have been a problem without precedent in engineering.

> *(j) The loss of life on August 29, 1907, might have been prevented by the exercise of better judgment on the part of those in responsible charge of the work for the Quebec Bridge and Railway Company and for the Phoenix Bridge Company.*

This is the only reference in the report to the loss of life, except for a note in Appendix 12, p. 95, titled "A Description of the Fallen Structure." "Out of eighty-six men on the work, only eleven escaped with their lives." Oscar Laberge, who survived the collapse but later died from his injuries, brought the total number of men killed to seventy-six.

The loss of life would certainly have been prevented if the work had been stopped and the men taken off the structure once the bends in chord A9-L were discovered on August 27th. Unfortunately, those in responsible charge were not at site; they were more than 500 miles away. None of the contractor's people at site could act without getting direction from the engineers in Phoenixville, and Hoare would not, and presumably, could not, make a decision without Cooper's advice. The Quebec Bridge and Railway Company and the Phoenix Bridge Company deprived their people of the power to deal with such emergencies. The federal government, who was financing the venture, had no one at site, at Cooper's insistence.

The commissioners wrote, "It was clear that on that day the greatest bridge in the world was being built without there being a single man within reach who by experience, knowledge and ability was competent to deal with the crisis. Mr. Yenser was an able superintendent, but he was in no way qualified to deal with the question that had arisen. Mr. Birks, well-trained and clear headed, lacked the experience that teaches a man to properly value facts and conditions. Mr. Hoare, conscious that he was not qualified to give judgment, simply assented to the courses of action that had been determined on by Messrs. Yenser and Kinloch, and made no endeavour to make a personal examination of the suspected chords."[15]

Engineering News summed it up this way: "Real authority lay at New York, 699

THE FINDINGS OF THE COMMISSION

miles away, with an engineer who had never seen the structure for which he was actually carrying the entire engineering responsibility."[16]

Although the commissioners were satisfied that no one connected with the work was expecting immediate disaster, they pointed their finger at Hoare. "Mr. Hoare was the only senior engineer who was able to reach the structure between August 27 and August 29. He was fully advised of the facts yet did not order Yenser to discontinue erection, which he had power to do. We consider that he was in a much better position than any other responsible official to fully realize the events that had occurred, and his failure to take action must be attributed to indecision and to a habit of relying upon Cooper for instructions."[17]

Cooper and Deans knew that Hoare would not act on his own; it was up to either of them to stop the work and get those men off the bridge. Cooper at least took a step in that direction with his poorly worded telegram to Deans, but Deans ignored it. It can be argued that it was Deans who, just hours before the collapse, had the last clear chance to save the lives of those men. He read Cooper's telegram at three o'clock. He had spoken to the site that morning and was aware that they had moved the small traveller out the previous day and were at that moment adding more steel to the structure. He knew all of this, yet he ignored Cooper's instruction to add no more load to the bridge. Although it can be argued that Cooper's telegram was not clear, Deans made no effort to seek clarification from him, nor did he make any attempt to contact Hoare or anyone else at the site. Instead, he told his people to attend a meeting with McLure at five o'clock. As that meeting ended, the bridge collapsed.

All three Commissioners were actively involved in questioning the witnesses throughout the inquiry; Deans appeared three times in Quebec and was questioned twice in Phoenixville. At no time was he asked by any of the commissioners why he didn't act on Cooper's instructions, why he didn't order the work stopped, or at least communicate with Hoare to get his instructions. None of the lawyers raised it either, including the workers' lawyer, Davidson.

Cooper was asked about his telegram, why he sent it only to Deans and not to Hoare as well, as he had previously done on a matter of much less importance. Cooper told the Commission that he had decided "that the shortest and quickest method of reaching the bridge was through the Phoenix Bridge Company, who I knew had direct wire and telephonic communication with their office at the bridge. On the previous occasion when I had stopped work on the bridge by communicating with the chief engineer of the Quebec Bridge Company there was no emergency before me."

The only mention of Cooper's telegram in the Commission's report is in Appendix 11, p.93: "With reference to Cooper's telegram, Deans knew that he was in possession of later information from the bridge than had reached Cooper, and therefore decided to wait for McLure and, afterwards, for the arrival of Mr. Birks's letter of August 28th before taking action."

It is far from clear what "later information" Deans had that Cooper didn't. Deans had spoken to the site that morning when Birks told him he believed the bends were not recent. When Cooper arrived at his office on the morning of August 29th, McLure showed him the telegram that he had just received from Birks—it was the same information that Deans had received. The only information that Deans had that Cooper didn't have was the fact that Yenser was continuing with the erection of the suspended span, which should have, but apparently did not, raise an alarm with Deans.

The harshest criticism leveled at Deans came from Holgate in appendices 7 and 9. "Mr. Deans's actions in the month of August 1907, and his judgment, as shown by the correspondence and evidence, were lacking in caution, and show a failure to appreciate emergencies that arose."[18]

"We consider that Mr. Deans was lacking in judgment and in sense of responsibility when he approved of the action of Mr. Yenser in continuing erection, and when he told Mr. Birks and Mr. Hoare that the condition of the chords had not changed since they left Phoenixville."[19]

The Commission noted, "No evidence has been produced before the Commission in proof of the correctness of Mr. Deans's statement about the chords, and Mr. Szlapka's calculations showed that the rivets were even then loaded to their maximum specified stress of 18,000 pounds per square inch. The theory underlying these calculations is very questionable, but it was adopted in the design of the bridge, and we cannot understand why its warning was so entirely disregarded in the face of the consequences that might result."[20]

In spite of this, the Commission was "Satisfied that no one connected with the work was expecting immediate disaster, and we believe that in the case of Cooper his opinion was justified. He understood that erection was not proceeding, and without additional load, the bridge might have held for days."[21]

There was a discrepancy in the evidence given by Cooper and by McLure as to whether or not Cooper was aware that the work of erection was continuing at site. In his evidence before the Commission, McLure testified that he told Cooper, "It was my understanding when I left the bridge that there would be no more metal erected until advices were received regarding these chords."[22]

During his initial examination in New York, Cooper testified that on August 29th, McLure told him, "Mr. Cooper, they have moved out the small traveller, but we estimated that it will not add to the strain on chord 9 more than 50 pounds per square inch, but they are going on this morning to erect more of the work; do you think that is right?"[23]

Cooper testified that he told McLure, "By no means right."

During his second examination in New York, Cooper was asked, "Did you consider at noon on August 29 that the collapse of the bridge was imminent?"[24]

"I did not think at the time," Cooper testified, "that without additional loading the collapse was so imminent that a remedy could not be applied, but I was not aware at the time that they were adding new material and had been for the previous day."

The Commission did not address this discrepancy and appeared to disregard Cooper's initial testimony on this critical point. As far as the Commission was concerned, Cooper understood that the work of erection had stopped. The commissioners accepted his evidence that he was not expecting imminent disaster. Cooper had admitted, however, that the bent chords constituted an emergency; unfortunately, his telegram did not reflect this.

It should be noted that there was no evidence that McLure had any contact with the bridge after he left for New York at noon on Wednesday. As far as he and Cooper were aware, they had moved out the small traveller, but had stopped the erection of the suspended span and no more steel was being added. Thus, it's likely that McLure's evidence on this point is correct.

The young McLure was not completely exonerated by the Commission, however. The Quebec Bridge inspector, Kinloch, told the Commissioners that "It was understood that McLure would immediately wire me if Mr. Cooper took a serious view of the situation, but this he failed to do."[25]

McLure never mentioned this in any of his testimony before the Commission, nor was he asked about it during his four appearances before the Commission, yet the commissioners accepted Kinloch's evidence; the report states, "Mr. McLure had promised to wire Mr. Cooper's decision to Mr. Kinloch immediately, but he did not do so."[26]

There was no mention at the coroner's inquest of McLure's alleged promise to wire the site. McLure told the jury about his meeting with Cooper in New York. "Mr. Cooper appeared to be slightly worried and seemed not to be of the opinion that it was dangerous but that it should be looked after. He never told me not to put anymore load on the bridge and he did not tell me either to telegraph any instructions to Quebec. I had instructions from Mr. Hoare to go to Phoenixville and Mr. Cooper told me also to go there."[27] If McLure had, in fact, made such a promise, he would no doubt have hesitated to send a wire on his own initiative after witnessing Cooper struggle with the decision of who to send his telegram to. This also raises the question of what Kinloch would have done after receiving such a telegram, and whether or not Yenser and Birks would have acted on their own and ordered the men off the bridge.

The commissioners accepted Kinloch's testimony without giving McLure a chance to confirm or deny whether he made such a promise, and if he did, why he didn't follow through. No doubt, the words of the commissioners would have weighed heavily on the unfortunate young man.

Birks, more than anyone, had put his faith in the senior engineers. He simply could not accept the fact that the bridge was failing, and he convinced himself and others that the bends had been there when the members went into the bridge. Deans described Birks to the commissioners: "A statement of Mr. Birks's experience in no sense conveys a proper estimate of his ability, which was unusual for a man of his years. He was specially fitted by character and temperament for the work entrusted to him. He was fearless—able to climb over the entire structure. He had a lovable character and that about him which instantly demanded respect and confidence. He could have his orders carried out readily without friction. It would be difficult to find a man combining the many traits of mind and character which so eminently fitted him for the position of resident engineer of erection." No doubt the young engineer would have been persuasive.[28]

The commissioners also defended him. "Birks knew better than anyone else on the work the care with which the calculations and designs had been made. He was familiar with the experience and abilities of the designers and could calculate that the stresses were far below the expected maximum. To engineers, the force of such reasoning is very great, and we do not consider that the confidence Mr. Birks placed in his superiors was in any way unusual or unreasonable. There was no misunderstanding, however, on his part. He realized that if the bends had not been in the chord before it was erected, the bridge was doomed, and although Mr. McLure had evidence that the bends had increased more than one inch in the course of a week, and although Mr. Kinloch was positive that the bends had very recently greatly increased, and although Mr. Clark stubbornly maintained that the chord was absolutely straight when it left the Chaudière yard, Mr. Birks still strove to convince himself that they must have been mistaken. Mr. Hoare evidently concluded that the matter was too serious for him to settle by any offhand decision and approved Mr. McLure's mission to New York, wisely requiring that he should get all possible facts before leaving, so that Mr. Cooper need not wait for further information on which to base a decision."[29]

An important factor that was not raised by the Commission was the very real risk that a work stoppage in late August might have resulted in the loss of the remainder of the construction season. Phoenix had planned to work until November and their goals for the 1907 season were already in jeopardy. There had been unrest among the workers, and a shutdown to address the problem with the chords would no doubt have made matters worse. In that case, many of the workers would most likely have left. It would have been difficult for someone in authority to make the decision to call the men off the structure, thereby potentially pulling the plug on the season. Hoare didn't want to do it, and neither did Cooper. There may have been financial consequences to the Quebec Bridge Company for doing so if Phoenix could show they were not at fault. Deans understood what Cooper's telegram meant, but he wanted so much to believe his young protégé, Birks. He simply was not going to risk losing the rest of the season until he had read Birks's letter, which was expected the following morning. The undue pressure to complete the work by 1908 had been brought to bear by the government of Canada and by the Quebec Bridge Company.

The Commission was convinced that no one was expecting imminent danger. It's difficult to argue to the contrary. Everyone, it seems, thought the bridge would hold out for a few days, at least. Kinloch, who discovered the bent chord, was on the bridge fifteen minutes before it collapsed, Yenser went down with the big traveller, and Birks was on a car behind the locomotive when it went over the end of the span. Even Haley, the union steward who fell four hundred feet into the water, testified that he thought he would get a chance to look at the chords that night. Szlapka collapsed when he heard the news. The bridge was so massive in its dimensions that no one thought it could possibly fail, but it did, and it claimed seventy-six lives.

> (k) *The failure on the part of the Quebec Bridge and Railway Company to appoint an experienced bridge engineer to the position of chief engineer was a mistake. This resulted in a loose and inefficient supervision of all parts of the work on the part of the Quebec Bridge and Railway Company.*

This is the only finding that admonishes any of the organizations involved, and it is narrow in its scope, to say the least. Although the Commission characterizes the conduct of the Quebec Bridge Company as a "mistake," their appendix states that they could have shown "better judgment."

The Commission had this to say about Hoare: "Mr. Hoare had a high reputation for integrity, good judgment, and devotion to duty. From the standpoint of personal character and knowledge of Quebec and its people, no better man could have been found, and the evidence throughout shows that to the best of his ability the Company was faithfully served. There is, however, nothing in Mr. Hoare's record that would indicate that he had the technical knowledge to direct the work in all of its branches.

"The Company's directors do not seem to have realized the importance of the duties pertaining to Mr. Hoare's position, and while believing that he was not competent to control the work, they still gave him the position, the powers, and emoluments of the office of chief engineer."[30]

The report continues, "It is our opinion that the Quebec Bridge Company would have shown better judgment had it employed a larger staff under the direction of an independent man of wider technical knowledge and who would have been sufficiently forceful to hold his own against the contractors."[31]

The failure of the federal government to appoint its own inspector did not go unnoticed, at least by the press. In an editorial, *The Gazette* (Montreal) condemned the Quebec Bridge Company for its failure to appoint an experienced engineer, but then turned its attention to the government. "The bridge fell and four score men perished. The same incapacity to appreciate its duty marked the conduct of the Dominion Government. It was providing nearly seven million of money for use by

THE FINDINGS OF THE COMMISSION

a company that by its own investment did not show confidence in its enterprise. It was as disregardful of its duties as was its protégé. It allowed professional punctilio on the part of a man who had no relations with it to prevail over the intention it seems to have once entertained of appointing an inspector in its own and the country's behalf. It stands today indicted by the report of its own commission of incapacity to properly fulfill its functions. The country knows the result."[32]

> *(l) The work done by the Phoenix Bridge Company in making the detail drawings and in planning and carrying out the erection, and by the Phoenix Iron Company in fabricating the material, was good, and the steel used was of good quality. The serious defects were fundamental errors in design.*

> *(m) No one connected with the general designing fully appreciated the magnitude of the work nor the insufficiency of the data upon which they were depending. The special experimental studies and investigations that were required to confirm the judgment of the designers were not made.*

Prior to the work commencing, the president of the Phoenix Bridge Company told his people to carry out any tests they, or the consulting engineer, deemed necessary. Phoenix owned all of the necessary testing equipment, including the largest compression testing machine in the world. Careful thought as to the designing of the bridge members was given only to the tension eyebars, and extensive testing was carried out only on the eyebars. Cooper presented a paper on eyebars at a meeting of his peers. Little thought was given to the compression chords, and no tests of any kind were performed on them until after the collapse. In 1908, Phoenix, on its own initiative, built a one-third scale model of the compression chords, and tested them on their machine. Of the seven hundred eyebars that were in the wreckage, only one was found to have snapped.

Clearly such tests could have been carried out during the design phase, but this would have taken valuable time. The rush to keep its manufacturing facilities busy was, as always for this Company, of paramount importance.

(n) The professional knowledge of the present day concerning the action of steel columns under load is not sufficient to enable engineers to economically design such structures as the Quebec Bridge. A bridge of the adopted span that will unquestionably be safe can be built, but in the present state of professional knowledge a considerably larger amount of metal would have to be used than might be required if our knowledge were more exact.

R.C. Douglas, the chief bridge engineer for the Department of Railways and Canals, was sixty-two years old and had thirty-seven years with the department. He testified at the inquiry but said very little. In July 1908, he too appeared before the Commons Special Committee looking into the financial aspects of the collapsed bridge. Douglas testified that had the unit stresses that he recommended been adopted, the structure would have been thirty-five percent heavier and likely have been successful. Also, the chords would have had continuous cover plates, at least on one side, instead of the lattices. Douglas's assertions were proven out by the tests conducted by Phoenix on the smaller-scale versions of the compression chords.[33]

(o) The professional record of Mr. Cooper was such that his selection for the authoritative position that he occupied was warranted, and complete confidence that was placed in his judgment by the officials of the Dominion Government, the Quebec Bridge and Railway Company, and the Phoenix Bridge Company was deserved.

This last finding of the Commission absolves everyone but Cooper, including the three organizations involved. *Engineering News* referred to Cooper as "the nestor of American bridge designers." In the Commission's view, everyone involved was justified in placing their faith and reliance entirely on Cooper.[34]

The Commission wrote, "In the extent of his experience and in reputation for integrity, professional judgment, and acumen, Mr. Cooper had few equals on this continent, and his appointment would have been generally approved. His professional standing was so high that his appointment left no further anxiety about the outcome in the minds of all most closely concerned. As the event proved, his connection with the work produced in general a false feeling of security."[35]

The commissioners noted, however, "The impression left with us is that throughout the work Mr. Cooper was in the position of a man forced in the interests of the work to take responsibility which did not fully belong to his position, and which he was not authorized to take, and that he avoided the assumption of authority whenever possible." When it came to his authority to change the specifications, though, Cooper demanded, and was given, an absolutely free hand. This fact, and its acceptance by the federal government, were not mentioned by the Commissioners.[36]

Regarding Cooper's inability to visit the site, and his reliance on the young McLure, the commissioners noted, "We are at a loss to understand why Mr. Cooper under the circumstances did not place a more experienced man in full local charge of the inspection of erection. We must recognize, however, that the power of making such an appointment did not rest with Mr. Cooper, and that Mr. Hoare has stated in evidence his conviction of his own ability to handle the work." Once again, the commissioners neglect to raise the fact that it was Cooper who insisted, over the objections of Hoare, that McLure be appointed as the Company's inspector. Nor is there mention of Cooper's absolute refusal to allow a peer review of his work, as proposed by the Department of Railways and Canals' chief bridge engineer and his boss, the deputy minister, Schreiber. There can be little doubt that the presence of such an engineer at site, representing the government's interests, would have benefited, and even perhaps saved, the project.[37]

The Engineering Record had this to say about Cooper's laments: "Perhaps the most painful part of the evidence is that in which the Consulting Engineer makes the plea of impaired health for not exacting from both the contractor and the Quebec Bridge Company certain requirements of design and plans in the one case, and the necessary organization for the proper performance of the work on the other. Unfortunately, such pleas are admissions of official shortcoming; however much a man may feel the disability of ill health, they give him no relief from official responsibility. There is one only clear way by which he can divest himself of the responsibilities of official position, and that is by a formal withdrawal from it…. The Consulting Engineer makes a further point in his evidence that the fee he received was quite insufficient to enable him to maintain a proper office work force for the discharge of the duties imposed upon him in his official capacity…. When he accepted the fee, he accepted all of the responsibilities of the position. No engineer has any right whatever to consider his responsibilities lessened because his fee is not as large as it should be…. The failure of the Quebec Bridge reflects in no way whatever upon the American engineering profession, it simply shows that the exactions of responsibility unfortunately make no compromise with the disabilities of age and ill health, even when combined with a meager compensation."[38]

When it was all over, Cooper had been paid $32,225.[39]

The report included over a thousand photographs of the bridge during construction, and the wreckage. The commissioners wrote, "No description in words can give as correct an idea of the wreckage as will be obtained by a study of the photographs in exhibit 34. The principal feature to be noted is the comparatively uninjured condition of all members except some of the lower chords, posts and sub-posts which, by reason of their position, had to bear the larger portion of the forces developed by the fall, and completely failed under them.

"All connections except the splices of the lower chords proved to be as strong as—and in most cases, much stronger than—the body of some of the members they connected; the compression members and their splices have shown themselves to be the weakest parts of the structure."

The cost of the inquiry totaled $53,500.[40]

CHAPTER 23

THE AFTERMATH

N OCTOBER 1908, PRIME MINISTER Wilfred Laurier received a letter from John McMahon of the Overland Securities Company in Denver, Colorado. McMahon, a native of Canada, attached an article that had recently appeared in the *Toronto Globe*, regarding a speech delivered by Lomer Gouin, premier of Quebec. In his remarks, the premier told his audience that the great catastrophe of the Quebec Bridge the previous year, when so many brave men went down to their deaths, weighed heavily on his heart. But, he pointed out, "Much as it was to be deplored, it was purely and simply an accident, for which the government could not be held responsible. They had placed the matter in the hands of the most eminent of engineers, and, as sometimes happens, the latter had blundered. The precious lives could not be recalled, but the bridge, which was a national necessity, would certainly be rebuilt." The *Globe* noted that the premier's remarks were punctuated throughout with loud applause.[1]

In his letter to the prime minister, McMahon also recalled the Tay Bridge disaster in Scotland in 1879. The recently opened bridge had collapsed during a gale, taking a trainload of passengers with it, killing all seventy-five people on board. A board of inquiry attributed the failure to faulty design and construction. The design engineer, who also supervised the construction, had failed to allow for the extreme wind loads, which were not uncommon in that area. McMahon wrote, "The loss of that splendid bridge with a great train load of passengers was a terrible affair, yet among all of the many articles censuring those in charge of the construction of that structure I do not remember seeing a single word of adverse criticism against anyone other than the engineer in charge of construction. This same line of conduct should prevail regarding the unfortunate Quebec Bridge affair."

By laying the blame principally on technical errors by two engineers, Professor Kranakis points out that Holgate and his colleagues depoliticized the inquiry, since "It allowed divisive political questions to be sidestepped, such as whether the Canadian government had been irresponsible in not actively overseeing a project that involved substantial public investment. If the commission had directly linked the Phoenix Bridge Company policies to the disaster, it could have triggered the company's downfall, hurting the entire workforce as well as the parent company, Phoenix Iron, which had performed its work successfully."[2]

The Commission's failure to address the organizational context of the disaster was addressed in an *Engineering News* editorial on September 17, 1908. "What in some respects is the most weighty feature of that report is the picture it gives of the heterogeneous, uncentralized, and non-responsible organization of the Quebec Bridge Company; we might say, the utter absence of authoritative, purposeful engineering direction. The underlying causes of this state of affairs are hardly even hinted at in the report, and they need not concern us now. The facts themselves are clearly enough demonstrated again and again; we may take as sufficient proof the bald statement that the huge undertaking of erection was going on without the presence of an engineer-in-charge."[3]

After the report of the Commission was tabled in March 1908, the Liberal government appointed a Select Committee of the House of Commons to "investigate the conditions and guarantees under which the Dominion Government paid moneys to the Quebec Bridge and Railway Company, and endorsed or guaranteed its bonds, and what measures were adopted by the government to ensure the preparation of suitable plans of construction and the proper execution of the same, and what security the government at present possesses for the sums already received by and guarantees given to the Company."

The president of the Quebec Bridge Company, Parent, appeared before the Select Committee, and testified that he had received legal advice that the Quebec Bridge and Railway Company was not responsible for the loss of life or any property damage. The workers were the employees of the Phoenix Bridge Company,

THE AFTERMATH

which was liable for the accident. According to Parent and his legal advisers, the Quebec Bridge Company was not responsible for the accident.

When he was asked how much he thought it would cost to complete the bridge, Parent told the committee, "Naturally I am not an engineer, but it is my inmost conviction that you can rebuild the Quebec Bridge, the bridge properly speaking, for less than five million dollars." His statement as to his engineering qualifications was correct, at least. In spite of this, however, he had an "inmost conviction" that it could be done for that dollar amount. It is not unusual for such bureaucrats and politicians, who have no experience whatsoever with such large engineering projects, to hold very strong opinions as to how much they will cost. Almost invariably, they're wrong. The taxpayer continues to suffer, even today, from such foolishness.[4]

While the Select Committee hearings were taking place in Ottawa, the Quebec Bridge and Railway Company signed a supplemental agreement with the Phoenix Bridge Company on April 21, 1908, whereby Phoenix was to build the superstructure for the new bridge. It is not clear what involvement, if any, the Dominion Government had in any of this.

At the conclusion of the hearings before the Select Committee, two reports were prepared by the members of the Committee: Report B, which was drafted by the loyal opposition but not adopted, concluded that "The Government, wholly unrepresented upon the work, left the public interest absolutely in the hands of the Quebec Bridge Company which, in the opinion of your Committee, was incompetent and, having regard to the relations between it and the Government, utterly unfitted for that position.

"Your Committee are of the opinion that the Government stand without much, if any, useful recourse against the Quebec Bridge Company for the sums disbursed or for which the Government has rendered itself liable."

Report A, which was prepared by the ruling Liberals, was adopted, and expressed no opinion on the competence of the Quebec Bridge Company. The report was presented for adoption in the House of Commons on July 17, 1908, a few hours before the summer prorogation. The presenter, William Chisholm, Liberal MP from Antigonish, told the House, "Now that the bridge has fallen, people naturally try to find excuses and to throw blame on somebody, and in this case it is thrown on the Quebec Bridge Company with the view, perhaps, of making a little political capital against the government. But, if you look at it fairly and squarely, I submit there is nothing in the evidence or the circumstances which would justify the charge that there was any carelessness or negligence on the part of the government or of the Quebec Bridge Company. The minority report undertakes to throw blame on Mr. Hoare. If there is blame to be thrown on anybody it must be thrown on Mr. Cooper, but I do not think that anybody will say that Mr. Cooper was not recognized as an authority. It is true that his judgment failed in this case, but the greatest scientists in the world make mistakes, and their judgment is apt at times to be at fault." Both reports were tabled in the House of Commons.

The Globe reported on the discussion that took place in the House regarding the reports of the special committee. Speaking on behalf of the opposition, Mr. Robitaille charged S.N. Parent with "willful neglect, which extended almost to criminal negligence." Robitaille also asserted that the jury in the coroner's inquest was packed so as to assure a favorable verdict for the Quebec Bridge Company. Parent responded to the allegations with a vigorous defence of the bridge company.[5]

Having received the report of the Select Committee, the House of Commons quickly passed a bill to "Authorize the government of Canada pursuant to the agreement between it and the Quebec Bridge and Railway Company, dated the 19th day of October 1903, and set out in the Schedule to chapter 54 of the statutes of 1903, to exercise the right to take over the whole of the undertaking, assets, property, and franchises of the Quebec Bridge and Railway Company, upon the terms and conditions set forth in the said agreement and the said Act, and that the moneys required to pay therefore be appropriated for that purpose out of the Consolidated Revenue Fund." The Quebec Bridge and Railway Company would cease to exist on December 1, 1908, but prior to that, it was to be indemnified by the Dominion Government, as were its shareholders. The government's liability resulting from the failed venture totaled $6,818,913.34. The Canadian taxpayer would shoulder the financial burden resulting from the collapse, as well as the cost to build a new bridge, which was as yet unknown. It would no doubt be more than five million dollars.

Immediately after the Dominion Government stepped into the shoes of the Quebec Bridge Company, it launched a claim against Phoenix for all damages resulting from the fall of the bridge. Phoenix counter-claimed for what it alleged to be a balance of $293,000 still owing under the contract, as well as for reimbursement of the claims it had settled with its employees and other parties, totaling

THE AFTERMATH

$106,000. Phoenix also claimed that under the terms of the supplemental agreement, it was entitled to construct the new bridge.

The parties settled their dispute in March 1910, with Phoenix waiving its right to the outstanding amount owed, as well as to the unused materials stored near the site. It also waived its claim for reimbursement of the third party claims, and its right under the supplemental agreement to build the replacement bridge. Finally, Phoenix agreed to make a payment to the government of $100,000. Following the collapse, the Quebec Bridge Company had neglected to call upon the $100,000 performance bond that had been provided by Phoenix at the time the contract was entered into in 1903. The surety had refused to pay on the bond as the notice had not been delivered in the time specified. This amount formed part of the settlement. During the settlement negotiations, the government was made aware that the assets of the Phoenix companies were held by the parent, the Phoenix Iron Company. The Phoenix Bridge Company had next to no assets; the parent was protected. The mutual release was dated March 12, 1910, and it put an end to the Phoenix Bridge Company's involvement with the Quebec Bridge project.

When, in 1903, the Canadian government had concluded the financial arrangements to provide guarantees for the Quebec Bridge Company, one of the conditions imposed upon the company was to obtain additional, paid-up share subscriptions, in the amount of $200,000. The total amount of stock paid in was $263,700. When the government took over the company in 1908, the shareholders of the Company were held harmless for any and all liabilities arising from the collapse. This decision also resulted in the government purchasing the shares of the failed company at face value, plus five percent interest per annum, together with a bonus of ten percent. This was all in accordance with the 1903 agreement between the Quebec Bridge Company and the Dominion Government. The shareholders were paid $355,279; a total return of 35 per cent for a bridge that was at the bottom of the river.[6]

Henry Holgate, the chairman of the Royal Commission of Inquiry testified at length before the Select Committee. He recommended strongly that the second replacement bridge should be a public work and that three of the most prominent engineers in the world should be appointed and given absolute control over the project. Engineers from Britain, America, and Canada were appointed by the federal government to oversee the design and construction of the new bridge under the auspices of the federal Department of Railways and Canals. They were each paid $12,000 per year.

Engineering News noted, "There will be a more anxious, a more rigid, scrutinizing of the conclusions and the designs produced by the committee. This fact renders their work correspondingly more delicate and of greater responsibility, and therefore more difficult. Judgment of fine metal will be needed. May it not be corroded by the steam rising from the ever-boiling pot of Canadian politics! May the committee accomplish its task successfully and worthily."[7]

It took almost two years to clear the debris from the shore of the river; the wrecked cantilever and suspended spans remain at the bottom of the St. Lawrence. The second bridge took more than seven years to complete, at a cost of over fifteen million dollars, three times Parent's "inmost conviction" estimate. With a span of 1,800 feet, it is still the longest cantilever bridge in the world.

Prior to calling for tenders for the second bridge, the design for the superstructure was completed under the board of engineer's supervision, complete with detailed drawings, a process that took more than two years. The replacement bridge was designed to accommodate two rail lines, two roadways, and two pedestrian walkways. Several construction companies submitted bids, including the St. Lawrence Bridge Company, a partnership between the Dominion Bridge Company of Lachine, Quebec, and the Canadian Bridge Company of Walkerville, Ontario. In light of past events, the contract for the new bridge included the following clause:

> "The contractor must satisfy himself as to the sufficiency and suitability of the design, plans and specifications upon which the bridge is to be built, as the contractor will be required to guarantee the satisfactory erection and completion of the bridge, and it is expressly understood that he undertakes the entire responsibility, not only for the materials and construction of the bridge, but also for the design, calculations, plans and specifications and for the sufficiency of the bridge for the loads therein specified."

THE AFTERMATH

Since the St. Lawrence Bridge Company was a special entity created for the project, its parents, Dominion Bridge and Canadian Bridge, were required by the government to sign the contract, and become joint and several guarantors for the proper execution of the work.

St Lawrence submitted two bids, one based upon the board's design, and another modified design of its own, whereby the roadways were eliminated. The alternate design complied with the specifications set by the board except for the roadways, and reduced the cost from $11,246,100, to $8,650,000. The board recommended the St. Lawrence proposal to the government, and a letter of confirmation was sent to its president, Phelps Johnson, who brought more than forty years of bridge engineering experience to the project. Johnson developed the "K" truss design used for the bridge. He replied with his acceptance on behalf of the two partners on February 23, 1911, and a contract was signed on April 4, 1911.

Eleven years later, on January 25, 1922, Phelps Johnson, the project manager for the second bridge, would attend a meeting of the Engineering Institute of Canada in Montreal, where Professor Haultain would give a speech on "The Romance of Engineering." He was one of the seven past presidents of the EIC who listened to Haultain at the retiring president's dinner later that evening. Phelps Johnson would become one of the Seven Wardens of the Ritual of the Calling of an Engineer.

Although the two piers for the southern span came through the collapse largely unscathed, none of the piers on either side of the river could be used for the replacement bridge. The new bridge was twenty-one feet wider and the entire structure was relocated sixty-five feet south. The granite face-stone from the original piers was salvaged and re-used, however. The new piers were built by the same company that had built the original piers, M.P Davis & Co.

Charles Schneider, the consulting engineer who was tasked with examining the Phoenix design of the superstructure, recommended that the old design be abandoned, which meant that all of the steel that had been fabricated for the northern span had to be scrapped. The new design differed in many respects from the original. The cross-sectional area of the compression chords was more than twice that of the previous structure. The superstructure was seventy-three percent heavier than the Phoenix design, and utilized nickel steel, which was lighter and permitted higher unit stresses.

The second bridge suffered a tragedy as well. This structure also differed from the first in that the center span between the two cantilever arms was to be built separately, off-site. Whereas the central, or suspended span for the first bridge was to be built out from the end of the two cantilever arms one frame at a time, the designers of the second bridge decided that they could save a year on the schedule by building the 640-foot-long center span three miles downstream of the main site, in the shallow waters of Sillery Cove, where the riverbed is exposed at low tide. Once the cantilever arms were completed, the span would be floated to the site on barges and lifted into place with the use of hydraulic jacks.

The 5,000-ton center span was floated to the site on a barge on September 11, 1916. It was to be hoisted into place with specially designed jacks and fastened to the two cantilever arms of the bridge. At 7:40 a.m., the lifting hangers for all four corners had been connected, and at 8:50 a.m., the jacks began the 150-foot lift of the center span to the almost completed structure above. Thousands of spectators watched as the last piece of the longest span bridge in the world was finally being put into place. Hundreds of people stood on the north and south cantilever arms, most others watched from land, and many from boats. Eminent engineers from all over the world had travelled to Quebec to witness the feat. Several of these prominent engineers had arranged to witness the operation firsthand as passengers on the center span while it was floated from Sillery Cove to the bridge site.

At 10:00 a.m. the hoisting operation was suspended to give the workmen a break. The operation resumed at 10:30 a.m. At 10:50 a.m., the center span was fifteen feet above the barge. Suddenly, a loud noise, like the sound of a cannon, was heard. A temporary cruciform steel bearing casting had failed, dropping a corner of the span. Everyone watched helplessly as the southwest corner of the center span slid off its supporting lifting girder. With one corner free, the structure twisted and pulled the southeast corner off its support. Freed from the weight of the center span, the south cantilever arm sprang upward eighteen inches. Workers on the temporary wooden platform, at the outer end of the cantilever arm where the jacks were located, jumped or were thrown into the river 150 feet below as the violent movement shattered the platforms. Spectators on the cantilever arm were thrown violently into the air. Panic set in—*Oh no, not again!* People screamed and scrambled to escape as the vibrations and oscillations caused by the release of the center span went on. The two north corner supports held stubbornly for several seconds, but when the south end of the span had sunk thirty feet into the water, the north corner supports finally gave way. The entire span disappeared into the St. Lawrence—it was all over in a matter of seconds.[8]

Of the eighty men involved with the lifting operation, thirteen were lost and fourteen were injured. Many of them had been on the center span as it was being lifted into place. The visiting engineers were no longer on the structure; they had disembarked during the morning break.

The collapsed center span rests on the bottom of the St. Lawrence River, two hundred feet below the surface, next to the south suspended span and cantilever arm of the first failed bridge.

Undaunted, the engineers made another attempt, duplicating the design. A year later, the second center span was jacked into place and the bridge was finally completed in 1917.

Following the collapse of the first bridge, Cooper's career ended, his reputation in ruins. He lived the rest of his life in obscurity in New York. He died of pneumonia on August 24, 1919, at the age of eighty. Cooper was unmarried. The following day the *New York Times* obituary described him as "A builder of great bridges." The caption stated that he "foresaw the Quebec disaster" and that "His warning message would have saved lives." The article read, "When the first cantilever bridge across the St. Lawrence River at Quebec collapsed in 1907 and nearly 100 lives were lost, it was said the lives of all the men might have been saved had a telegram sent by Mr. Cooper been received and heeded. Mr. Cooper had been informed by an inspector of the dangerous condition of the bridge and had sent a telegram of warning, which was delayed in transit."[9]

That was the *New York Times'* take on Cooper's involvement; he was, in their view, a hero. Just two days before Cooper died, the Prince of Wales presided at the formal dedication of the second, replacement cantilever bridge at Quebec, which had been designed and built by others—the bridge that Cooper foresaw as his crowning achievement.

Edward Hoare went to work for the National Transcontinental Railway Commission, whose president was his old boss, Simon-Napoleon Parent. Hoare returned to the type of work he was well-suited for, the location and construction of Canada's new transcontinental railway system. Parent himself had been appointed to the position by his old friend and colleague, Prime Minister Wilfred

Laurier. Laurier's Liberals were re-elected in the 1908 federal election but suffered a bitter defeat in 1911 when Laurier called a general election to settle the issue of trade reciprocity with the U.S., a wish the Liberals had, dating back to 1891. Having governed Canada for fifteen years, Laurier became an energetic leader of the opposition. He died on February 17, 1919.

Peter Szlapka granted an interview to the Toronto *Globe* newspaper on March 11th, 1908, two days after the Commission's report was tabled in the House of Commons. He told the reporter that Phoenix had no criticism to make of the Commission's findings, but he claimed that he had, in fact, criticized the compression chord design that had been prepared under his direction, a design based upon Cooper's own specifications. When the drawings came back approved by Cooper, Szlapka said that he was still not satisfied with them and returned them to Cooper, calling attention to the lattice angles. According to Szlapka, Cooper's reply was "They are all right. Don't alter." He did not explain why these specific details were not brought out when he gave his testimony before the Commission.

The Globe noted the above statements by Szlapka, and in an editorial wrote, "As the Commission reports that the material put into the bridge was good and that the work was well done, it seems evident enough that the responsibility for the collapse must rest on Mr. Cooper. All this places him in a very unenviable position, but there seems to be no way of avoiding a disastrous close of an honorable professional record. In the light of these demonstrations and admissions only the most rabid partisan would think of holding either the Quebec Bridge Company or the Dominion Government responsible for the calamity both had done their utmost to obviate." There was no mention or recognition of the roles each of these organizations played in giving Cooper supreme authority.

Peter Szlapka continued in his engineering position with Phoenix but resigned some years later to take a position as a structural engineer in the construction of subways in Philadelphia. In 1933, he retired to the house he had built for his family in Phoenixville. He died in 1943, just a few days before his ninety-first birthday.[10]

John Sterling Deans continued as chief engineer for Phoenix for ten more years. His duties included the construction of the Manhattan Bridge which was, at the time, the heaviest suspension bridge ever built. He became vice president of the company in 1915 and died at his home in Phoenixville in December 1918.

THE AFTERMATH

The Phoenix Company survived the blow to its reputation; insurance covered its direct losses and it had a strong backlog of business. In 1910, Phoenix completed the Manhattan suspension bridge, which had a clear span of 1,470 feet, a triumph for the company. Phoenix survived for another half-century, ultimately closing in 1962. The firm ultimately built some 4,200 bridges, as far away as Chile, Russia, and China. After the collapse of the Quebec Bridge, however, the Reeves brothers shied away from big projects. Never again did they take on anything as heroic as the bridge over the St. Lawrence.

The average age of those killed in the 1907 collapse was twenty-eight; the youngest was fourteen-year-old Stanley Wilson from St. Romuald, and the oldest was forty-eight-year-old Tom Jocks, a widower from Kahnawake. Forty-four of the seventy-six victims were married, leaving behind seventy-seven orphans.

Several lawsuits were filed on behalf of the families against the Phoenix Bridge Company, including an action for $15,000 by the father of Zephirin Lafrance, in whose name the inquest was held. Claims from $1,000 to $3,000 were also made by the widows of the workmen from the Quebec area. By September 1908, all of the suits and claims against Phoenix were settled by Waitneight, the Phoenix timekeeper, and their lawyer, Stuart. In a summary report to Phoenixville, Waitneight wrote, "I desire to call your particular attention to the attached statement which shows that all who have suffered equally have been given practically the same amount, that is, widow, $750; widow and one child, $1,000; widow and two children, $1,100; widow and three children, $1,200; and the families that have lost one or more sons, $450 for each son." The wife and child of Oscar Laberge, who died from his injuries after the collapse, was settled by Phoenix for $1,200. The widow and seven children of Harry French received $1,800.[11]

Indian agent James Macrae had been appointed guardian of the fifty-two minor Kahnawake children left fatherless by the disaster. In September 1908, he accepted $100,000 in settlement for their claims.

Immediately following the disaster, a victims' fund was set up for the families of the workers who lived near Quebec City. Almost twelve thousand dollars was collected and distributed to the families. Nearly all of the men belonged to the ironworkers' union, which insured its members for a work-related death; the families promptly received a $100 death benefit. Phoenix covered funeral expenses.

Most of the married men from Kahnawake had been prudent about taking out life insurance with the Independent Order of Foresters; their wives received payments totaling more than $19,000.

Far from deterring the young men from Kahnawake, the "disaster," as it is commonly referred to by the natives, made high steel work that much more interesting. They were proud to be able to do such dangerous work, and they were admired on the reservation. The disaster was a terrible blow to the women, however. They took up a collection and erected a crucifix over the altar at the St. Francis Church in their village. Then, they came together and decided that never again would all of their men work on a single job. Referred to as "booming out," the men spread out across the country to find work, eventually making their way to the U.S. where they worked alongside Newfoundlanders, who were affectionately called "Fish," men who came from "around the bay," places like Conception Bay, Harbour Grace and Bacon Cove.

The day after the collapse of the bridge on August 29, 1907, the New York weekly magazine *Engineering News* (which later became the *Engineering News Record*) dispatched a team of journalists and photographers to Quebec to report on the collapse and the events that followed. From September 5, 1907, to March 19, 1908, the magazine provided the most extensive coverage of any news agency; pages of important testimony appeared weekly. The Report by the Commission tabled on March 9th was thoroughly reviewed and discussed. In its editorial of March 19, 1908, the magazine posited that the lesson of the Quebec Bridge is the contrast between the practical man, the man whose only training was the training of the shop and the field, and the engineer with a thorough technical education. "It need hardly be said," the editor wrote, "that we do not draw attention to this matter as an argument against college training for engineers. No man need be one whit less practical as an engineer for a four-years course in an engineering school. The real lesson to be drawn is the lesson of humility. Let us never undervalue the experience of the man who actually handles and works the materials. Better yet, it is the duty of the engineer to be even more practical than the workman. It is of course true that things like this have been said many times before; but the trouble is, we have not taken them to heart. We have supposed that they referred to the men who try to do professional work with nothing but theoretical knowledge. It

THE AFTERMATH

has not occurred to us that men in the top ranks of the profession, who have been building great engineering works for nearly a lifetime, needed such admonitions.

"And yet that is what the event shows. We, all of us, juniors and seniors alike, need to know more—to test our theories constantly in the light of new knowledge, to welcome such knowledge when it comes, well attested, from any source. Yes, surely, the great lesson of this greatest disaster is the lesson of humility."

The Quebec Bridge was designated an International Historic Civil Engineering Landmark by the Canadian Society for Civil Engineering and the American Society of Civil Engineers in 1987, and a Canadian National Historic Site in 1995. It still ranks as the longest span cantilever bridge in the world. It will probably never lose this designation as metal truss bridges and cantilever bridges have now been replaced by cable stayed bridges and pre-stressed concrete bridges. The Quebec Bridge has gone through several configurations of traffic; today, it accommodates three highway lanes, one rail line, and a pedestrian walkway.

With its tragic past and eventual triumph, the Quebec Bridge stands as a monument to our enduring Canadian heritage, and to our perseverance.

The bridge took two attempts and eighteen years to complete, from October 2nd, 1900, to August 21st, 1918; the total cost was $22,534,894.66[12].

Records indicate that ninety-four men died during its construction. The Phoenix chief engineer testified at the inquiry that five men died prior to the 1907 collapse. According to Deans, these five fatalities were attributable to the fault of the workers themselves. The week before the collapse, on August 20, 1907, Joe Ward fell 170 feet when he was trying to remove a pin from a shackle on the suspended span; of course, he wasn't tied off. Mandatory fall-protection would not come into effect until 1970.

Seventy-five men died during the collapse of the southern span, with one of the survivors dying later, bringing the total to seventy-six. On September 11, 1916, thirteen men were killed and fourteen were injured when a casting on the center span of the second bridge failed during the hoisting operation. The board of three engineers appointed by the federal government to oversee the design and construction of the second bridge prepared a final project report in which they lamented the loss of life during the second attempt and provided a casualty list. There is no other reference to safety in the report.

Today, we place much, much more emphasis on safety. A contemporary job report would begin and end with safety, reflecting the value we place on the well-being of our workers. Thankfully, we've learned that lesson well.

EPILOGUE

IT WAS NOT MY INTENTION to lecture on morals and ethics; I'm less than qualified to do that. My goal was to bring this true story to life, in a way that you will believe it and remember it. Hopefully all of us, professionals and laymen alike, will learn something from this tragic event in Canadian history. And if you are an engineer, may it guide you in your choices, and serve as a reminder of the Obligation that we took, on Cold Iron.

Ninety-four men died bridging the St. Lawrence River near Quebec City. Just days before the collapse in 1907, thirteen members of the Kahnawake Lacrosse team posed for this photo after a practice game. They were all ironworkers, mostly riveters. The bridge looms in the background. Only five of these men survived. (Kanien'kehaka Raotitiohkwa Cultural Center)

GLOSSARY

(a) The Quebec Bridge People

BARNES, John Hampton: General Counsel for the Phoenix Bridge Company; appeared briefly at the inquiry.

BARTHE, Ulrich: Secretary and member of the board of directors of the Quebec Bridge Company; he was also a member of the Quebec City council with S.N. Parent, who convinced him to join the board.

BEAUVAIS, Alexander: Foreman of a four-gang of riveters, all from Kahnawake; hired by Phoenix in 1905. His supervisor was C.E. (Slim) Meredith, the rivet boss. He survived the collapse.

BIRKS, Arthur: Resident engineer of erection for the Phoenix Bridge Company. Recent graduate of MIT, he was hired by Phoenix in 1904 and sent to the site in late 1905, after chord A9-L was dropped and repaired. He was appointed against Cooper's advice, who thought him inexperienced. Birks was convinced the chords were bent when they went into the bridge. He died in the collapse.

BRITTON, Ed: Electrician for the Phoenix Bridge Company; he worked on all parts of the bridge and the storage yards. He was always on hand when the travellers were to be moved.

CLARK, Horace: Phoenix foreman in the storage yard; he received and unloaded the metal, and checked it and stored it until it was needed at the bridge site. He was the foreman in the yard when chord A9-L was dropped, and he also supervised its repairs. He was certain the chord was straight when it left the yard.

COOPER, Theodore: Consulting engineer for the Quebec Bridge Company and a leading American bridge designer from New York. He insisted on having the final word on the design and specifications for the bridge. He was unable to travel to site for medical reasons, and so he never saw the structure for which he was carrying the entire engineering responsibility.

CUDWORTH, Frank: Resident engineer in charge of instrument/survey work for Phoenix; his duties also included taking photos.

DAVIDSON, W.H.: Lawyer representing the workers' unions at the inquest and the inquiry.

DAVIS, Charles: Workman; survived the collapse.

DAVIS, M.P.: Contractor for the substructure piers.

DEANS, John Sterling: Chief engineer for the Phoenix Bridge Company. Twenty years earlier, while Hoare was chief engineer for another railway company, Phoenix and Deans had designed and built bridges for his railroad. Deans set up a scheme whereby all decisions regarding erection were made by the engineers in Phoenixville. He ignored Cooper's instruction to add no more load to the bridge.

DOUGLAS, Robert: Chief bridge engineer for the Department of Railways and Canals since 1893. The department's standard specifications were founded upon general specifications drafted by Douglas in 1896. Neither he nor his department provided oversight for the superstructure since Cooper had the final say.

EDWARDS, E.L.: Inspector of materials at the mills and shops in Phoenixville for the Quebec Bridge Company. Although he reported to Cooper, he remained an employee of the Phoenix Iron Works Company.

HALEY, D.B.: Bridge worker and president of the Union; he started on the job in June 1907 and worked on the small traveller. He and some fellow workers went to see the bent chords in the cantilever arm in the final days before the collapse. He survived the collapse.

HALL, Ingwall: Workman; survived the collapse.

HOARE, Edward A.: Chief engineer for the Quebec Bridge Company. Civil engineer trained in Britain with experience in railway construction. He was hired initially to examine and recommend potential sites for the bridge and then was appointed chief engineer. He deferred to Cooper on all but monetary matters.

HUOT, Joseph Adolphe: Timekeeper for the Phoenix Bridge Company; survived the collapse.

JOHNSON, James: Workman; survived the collapse.

JOLICOEUR, Dr. G.W.: Coroner for the province of Quebec; presided at the inquest.

KINLOCH, Robert: Appointed inspector of erection for the Quebec Bridge Company by Hoare in 1905. He had twenty years experience in bridge construction; he was not an engineer. He knew there was something wrong with the structure but none of the engineers would listen to him. He did not, however, expect imminent failure.

LABERGE, Oscar: Workman; survived the collapse, but died later from his injuries, bringing the death toll to seventy-six. He had been recently married and his wife gave birth to their child just before his death.

LAFRANCE, Zephirin: Workman who was killed in the collapse, and whose death was the subject of the inquest. His father, who had the same first name, testified at the inquest.

LAJEUNESSE, Delphis and Eugene: Brothers and bridge men from nearby Lévis; they had started in July 1907 as riveters, working for the rivet boss, Slim Meredith. They survived the collapse.

MCLURE, Norman: Appointed by Cooper as the chief inspector of erection for the Quebec Bridge Company, over Hoare's objections, who had already appointed Kinloch. McLure had recently graduated from Princeton in 1904 with a degree in civil engineering. Since Cooper was unable to travel, McLure became his eyes and ears at site. He reported to Cooper weekly and travelled to New York frequently to keep the consulting engineer up to date.

MEREDITH, C.L. (Slim): Rivet boss for the Phoenix Bridge Company from Columbus, Ohio; died in the collapse.

MILLIKEN, A.B.: Superintendent of erection for the Phoenix Bridge Company, reporting to the chief engineer, Deans. Milliken was responsible for all of the company's projects in the U.S. and Canada.

NANCE, J.J.: hoisting engine operator on the small traveller, working next to Haley; he survived the collapse.

NORRIS, Frank: the manager of the Phoenix Iron Works shop in Phoenixville.

PARENT, Simon-Napoleon: President of the Quebec Bridge and Railway Company and chairman of its board of directors (1897-1908). Parent and the company were exonerated by the Commission. He was the mayor of Quebec City (1894-1906) and premier of the Province of Quebec (1900-1905); he had close personal and political ties with the prime minister of Canada, Wilfred Laurier, who was also from Quebec. Laurier appointed him chairman of the federal Transcontinental Railway Commission (1905-1911).

REEVES, David: President of the Phoenix Bridge Company and of its parent, the Phoenix Iron Company. His brother William was the general superintendent.

SCHREIBER, Collingwood: Chief engineer and deputy minister of the federal Department of Railways and Canals. Schreiber retired in 1905, before the erection of the superstructure began. He allowed Cooper to have the final word regarding the specifications and design of the superstructure; his department's approval was a mere rubber stamp.

SCHNEIDER, Charles Conrad: a well-known bridge engineer from Philadelphia. He formed part of the Royal Commission of Inquiry as he was asked to opine on the Phoenix design and advise whether any of the materials already manufactured by Phoenix could be used in the new structure. He advised against re-use of all but the piers, however, even the piers were not re-used. It was Schneider that Cooper had recommended should take his place in 1905, when it became evident that he would be unable to travel to site. Schneider found the design woefully inadequate.

SZLAPKA, Peter: Chief design engineer for the Phoenix Bridge Company. He was in charge of the design of the superstructure, which was submitted to and approved by Cooper. The Royal Commission found his design of the lower compression chords to be defective.

STUART, Gustavus G.: Local counsel in Quebec representing the Phoenix Bridge Company during the inquest and the inquiry. Stuart and Waitneight settled all claims against Phoenix.

WAITNEIGHT, Mr.: The Phoenix Bridge Company timekeeper. He telephoned the news of the collapse to Phoenix and was their representative immediately following the disaster. He worked with their local outside counsel, Stuart to settle all claims against Phoenix.

WARD, Joe: Workman who had looked at the joint on the cantilever arm with Haley earlier in August. The week before the collapse, he lost his balance and fell 170 feet into the river. His body came to the surface after the bridge collapse.

WILSON, Percy: Workman; survived the collapse.

YENSER, Ben: Full-time supervisor at site in charge of erection for the Phoenix Bridge Company. He was a thirty-six-year-old bridge man, who had started with the company when he was sixteen and had been a general foreman with them for the past ten years. He died in the collapse.

GLOSSARY

(b) Technical Terms

abutment: a support at one end of a bridge that carries the load of the structure to the ground and usually holds in place the earth fill adjacent to the bridge.

arch: a curved structural form in which the member acts principally in compression, producing both vertical and horizontal forces at its support or abutments.

beam: a structural member, usually horizontal, which when loaded, acts principally in bending and shear.

bearing: a bridge bearing is a component of a bridge that provides a resting surface between bridge piers and the bridge deck. The purpose of a bearing is to allow controlled movement and thereby reduce the stresses involved.

cantilever: a structural form in which a member is free at one end and restrained, or counterbalanced, at the support end.

cantilever bridge: a bridge form based upon the cantilever principal. Typically, cantilever arms project toward the center of the span from main piers; they are counterbalanced by anchor arms between the main piers and the anchor piers at each end. A simple span suspended between the two cantilever arms completes the structure. The weight of the suspended span and the cantilever arms is counterbalanced by the weight of the anchor arms plus the steel embedded in the anchor piers.

chord: the longitudinal top or bottom members of a trussed structure, which carry the principal tension or compression forces.

compression: a force acting on a member that tends to shorten it—it is the opposite of tension.

compression chord: the bottom member of a trussed structure, which carries the compression forces.

dead load: the loading imposed upon a structure by its own weight. Often added to this is snow load.

elastic limit: maximum stress or force per unit area within a solid material that can arise before the onset of permanent deformation. When stresses up to the elastic limit are removed, the material resumes its original size and shape.

erection: the process of assembling or raising a structure.

eyebar: a flat steel bar used as a tension member in chords or web members. The steel bar has a large "eye" forged at each end, which is then drilled for the insertion of a connecting steel pin.

falsework: temporary staging utilizing wood or light steel frames to support a structure during erection of the main structure until it is complete and self-supporting.

gantry: a raised platform supported by towers or side frames, which runs on parallel tracks carrying a crane or derrick.

"K" truss: a truss form in which each panel of the truss is subdivided by two diagonal web members meeting at the midpoint of a vertical member. The idea of the K truss is to break up the vertical members into smaller sections. This is because the vertical members are in compression; the shorter a member is, the more it can resist buckling from compression.

latticing: a system of diagonal bars or angles used to connect two or more sections of a structural member to enable it to function as a unit.

live load: the external load imposed by traffic moving over a structure. Other external loadings considered in design include the braking forces of trains, wind loads, and the effects of contraction or expansion resulting from temperature changes.

member: the individual structural steel elements that make up the superstructure, such as chord A9-L. Each member is made of plates, angles, flanges, bars, etc.

panel: an individual frame or segment of a truss structure defined by the arrangement of web members.

pier: an intermediate support for a bridge structure.

pin connected: a form of bridge erection in which members are connected by steel pins driven through pin holes drilled in the members. This method provides relatively simple and rapid erection compared to riveted, bolted or welded connections.

plate girder: I-beams made up from separate structural steel plates, as opposed to a rolled single cross-section, which are welded, or in older bridges, bolted or riveted together to form the vertical web and horizontal flanges of the beam.

shear: a force acting across a structural member as opposed to forces acting at each end.

simple span: a structure spanning between two supports and unconstrained at either support.

specifications: a set of documented requirements or technical standards to be satisfied by a material, design, product, or service.

steel: a refined form of iron, alloyed with carbon, giving it greater strength and wear resistance than either cast or wrought iron. It is also much easier and cheaper to manufacture. Ordinary carbon steel is the usual form for structural steel; this was used for the first Quebec Bridge. Nickel alloy steel has greater strength and durability; this was used for the second bridge.

strain: the ratio of the change in length to the original length of a member under load.

stress: the intensity of loading on a member, usually expressed in terms of load per unit of cross-sectional area, such as pounds per square inch.

superstructure: the upper part of a bridge that carries the live load. The abutment, piers, and other support elements are referred to as the substructure.

suspension bridge: a bridge carried on cables or chains that are continuous between anchorages at each end of the structure and supported from intermediate towers.

tackle: the lifting apparatus, such as cables, blocks and pulleys that is used to hoist the structural members into place and hold them steady until they are fastened together with temporary bolts.

tension: a force acting on a member that tends to extend or elongate the member—it is the opposite of compression.

traveller: a movable structure equipped with hoisting equipment used in bridge erection. Some travellers are designed to operate along temporary supports at the floor level of the structure being erected, as was the case for the Quebec Bridge. Others travel along the top chords of a completed truss structure, erecting it as they go.

truss: a strong, rigid structure made up of a triangulated assembly of members designed to transmit loads in direct tension or compression with no bending. A bridge is made up of trusses on each side of the bridge deck, which are then joined together by lighter members above and below the deck.

web: there are two meanings; the first and most relevant for our purposes is the vertical section of a rolled beam or plate girder separating the top and bottom flanges. The bottom compression chords on the Quebec Bridge were comprised of four built-up plate girders, joined together by lattices and end plates. The second definition is the vertical and diagonal members between the top and bottom chords of a truss.

wrought iron: a form of iron with a very low carbon content that is tough, malleable, relatively soft, and much less brittle than cast iron.

NOTES

1. Professor Haultain's Request

(1) The Canadian Mining Hall of Fame (info@mininghalloffame.ca): Professor Herbert Haultain. (2) University of Toronto Archives (UTA) 1910-1978, Haultain fonds, B1972-0005/005. (3) The Canadian Mining Hall of Fame: Professor Herbert Haultain. (4) UTA, H.E.T. Haultain fonds, B1972-0005/005). (5) *MacLean's*, "God Bless the 'Girls In Green'!", Gertrude E.S. Pringle, Feb. 15, 1922. (6) UTA, H.E.T. Haultain fonds, "Ritual of the Calling of an Engineer" B1982-0023/018(01). (7) Ibid., B1982-0023/018(04). (8) Ibid., B1982-0023/018(02). Note that Kipling's reply refers to Haultain's letter of October 19th, but the copies of Haultain's letter in the files show the date as the 18th. There is also one copy of Haultain's letter where the 18 was changed to a 19. Also, see Haultain's reply to Kipling, dated Nov. 23, 1923, where Haultain writes that Kipling's package was waiting for him on his return from a fortnight's (fourteen days) absence, which would make it the 19th. It would appear therefore that Haultain's letter was dated October 19th.

2. Rudyard Kipling's Response

(1) Kipling, Rudyard. *The Irish Guards in the Great War*, (London: Macmillan, 1923). The Irish Guards had been Kipling's son's regiment. Eighteen-year-old Jack was killed on his second day on the battlefront in France during World War I. (2) Carrington, Charles. *Rudyard Kipling, His Life and Work*. London: Macmillan & Co. Ltd., 1955, 14-20. (3) Kipling, Rudyard. *Something of Myself, For My Friends Known and Unknown*. London: Macmillan, 1937, 52; Pietre-Stones. *Review of Freemasonry*. Oct. 2008. (4) Carrington, 337; see also "*Sappers*," "*The Ship that Found Herself*" and "*007*." (5) Carrington, 265. (6) *A Book of Words*,

NOTES

Macmillan, 1928. A collection of thirty-one speeches delivered by Kipling. McGill University, Montreal, October 12, 1907. (7) ibid. (8) Carrington, 399; Nobel Prize in Literature 1907 – presentation speech, Nobelprize.org. (9) Kipling, "*The Sons of Martha*"; *The Holy Bible*, Luke 10, 38-42. (10) Carrington, 416. (11) Hughes, James. *Those Who Passed Through: Unusual Visits to Unlikely Places*, New York History, 2010, 91 (2): 146–151. (12) Carrington, 470. (13) UTA, Haultain fonds, B1982-0023/001(02), 1925-1927. (14) Ibid., B1982-0023/010(02). (15) Ibid., B1982-0023/007(01); ref. notes by Ross, Nov. 15, 1924- ("These should be of purest iron, preferably prepared electrolytically, either direct from Canadian ore, or else from some historic piece of metal"). (16) Ibid., B1982-0023/003(02). (17) Ibid., B1982-0023/006(02). (18) Ibid. (19) Ibid., B1982-0023/003(04). (20) Ibid., B1982-0023/010(02). (21) Ibid. (21) Ibid. (22) Ibid. (23) Kemp, Sandra. *Kipling's Hidden Narratives*, 102-3. These notes are based on those written by Donald Mackenzie for the Oxford World's Classics edition of *Puck of Pook's Hill* and *Rewards and Fairies* (1995) with the kind permission of Oxford University Press. This short story was written by Kipling in 1906, and published in an American magazine in 1909, before being collected with other short stories and poems into a volume titled, *Rewards and Fairies* the following year. (24) UTA, B1982-0023/10 (12). (25) Ibid., B1982-0023/018 (03).

3. A Bridge at Quebec

(1) Middleton, William D. *The Bridge at Quebec*. Indiana University Press, 2001, p.8. (2) Ibid., 10-13; Royal Commission Report, "Quebec Bridge Inquiry," Sessional Papers, No. 154 (1908), (note that the Report is in three volumes: Volume I is the Report of the Commissioners; Volume II is the transcript of the testimony taken during the inquiry together with printed exhibits and photos (Volume III is not referred to); for ease of reference, Volume I will be referred to as the Report I, and Volume II as the Inquiry II), footnote 2 in ch. 3 refers to Report, I:12. (3) Middleton, 25-27. (4) Archives – Archdiocese of Quebec, DCB Mobile beta, Vol. XIV (Parent was mayor of Quebec City 1894-1906, premier of Quebec Oct. 1900-Mar 1905 and president of Quebec Bridge Company 1897-1908). (5) House of Commons Select Committee Appointed to Investigate the Conditions and Guarantees Under Which the Dominion Government Paid Monies to the

Quebec Bridge Company, 7-8 Edward VII, A. 1908 - Report, with Minutes of Proceedings (Ottawa, 1908) (hereinafter cited as "Select Committee Report"), 99. (6) Inquiry (n. 2 above), II:370. (7) Report (n. 2 above), I:14. (8) Inquiry (n. 2 above), II:343.

4. Chief Engineer

(1) Tarkov, John. *A Disaster in the Making*, American Heritage of Invention & Technology 1 (Spring 1986), 10-17. (2) Middleton, 63. (3) As was the case for the Tacoma Narrows bridge, which oscillated and collapsed in 1940; Tubby the dog was the only fatality. (4) Report (Ch. 3, n. 2 above), I:15-18. (5) Ibid., 40. (6) Ibid., 16. (7) Ibid., 24. (8) Inquiry (Ch. 3, n. 2 above) II:420. (9) Ibid., II:508. (10) *The Globe and Mail*, October 3, 1900. (11) Inquiry (Ch. 3, n. 2 above) II:354. (12) Ibid., II:408; Tarkov, 4.

5. Specifications and Financing

(1) Report, (Ch. 3, n. 2 above), I:40; Middleton, 189 (fn. 7). (2) Report, (Ch. 3, n. 2 above), I:28. (3) Ibid., 146. (4) Inquiry, (Ch. 3, n. 2 above), II:410. (5) Ibid., 410-411. (6) *Montreal Star*, September 1907. Note that the Firth of Forth Bridge is still standing and in use today. (7) Inquiry, (Ch. 3, n. 2 above), II:389, 446. (8) Ibid., 346. (9) Inquiry (Ch. 3, n. 2 above), II:457. (10) Ibid., 457-458. (11) Ibid., 336. (12) Ibid. (13) Report (Ch. 3, n. 2 above), I:42. (14) Ibid., 43. (15) Ibid., 43-44. (16) Inquiry (Ch. 3, n. 2 above), II:546-547. (17) Report (Ch. 3, n. 2 above), I:45. (18) Ibid. (19) Ibid., 20. (20) Ibid., 13; 3 Edward VII., chap. 177.

6. Design

(1) Middleton, 53-55. (2) Inquiry, (Ch. 3, n. 2 above), II:385. (3) Ibid., II:348; Report, I:50. (4) *Engineering News* vol. 57, January 17, 1907, p. 53. (5) Middleton, 55. (6) Ibid., 184. (7) Report (Ch. 3, n. 2 above), I:100. (8) Ibid., 77-78; Middleton, 55. (9) Select Committee Report, 196-197. (10) Inquiry (Ch. 3, n.

2 above), II:393. (11) Ibid., II:348. (12) Ibid., II:391. (13) Report (Ch. 3, n. 2 above), I:150; Middleton, 65. (14) Report (Ch. 3, n. 2 above), I:57. (15) Inquiry (Ch. 3, n. 2 above), II:411. (16) Report (Ch. 3, n. 2 above), I:9, 37. (17) Inquiry (Ch. 3, n. 2 above), II: 345. (18) Ibid., 392.

7. Inspection Regime

(1) Inquiry (Ch. 3, n. 2 above), II:348-350. (2) Ibid., 45. (3) Ibid., 41. (4) Ibid., 469. (5) Ibid., 34-35. (6) Ibid., 46. (7) Ibid., 374. (8) Ibid., 351-352, (9) Ibid., 60. (10) Ibid., 146-147. (11) Ibid., 400.

8. The Builders

(1) Inquiry (Ch. 3, n. 2 above), II:321. (2) Ibid., 406. (3) Middleton, 61. (4) Mitchell, Joseph. *New Yorker Magazine*, Sept. 17, 1949, p. 38; See also: Owen, Don. *High Steel*, NFB Documentary, 1966. (5) Blanchard, David. *MELUS* – The Journal of Ethnic Studies 11:2, p. 45; There were 546,000 field rivets required for the bridge; Report (Ch. 3, n. 2 above), I:76. (6) Wilson, Edmund. *Apologies to the Iroquois with a Study of The Mohawks in High Steel by Joseph Mitchell*, Syracuse University Press, 1992, p.14. (7) Reputed to be among the best in Canada, my maternal grandfather, Henry Boudreault, who was a heater from Ontario, could toss them up to seventy feet, although he maintained that his accuracy diminished after sixty feet. (8) Average wage was twenty-two cents per hour; riveters earned fifty cents per hour.

9. The Dropped Chord A9-L

(1) Middleton, 56. (2) Inquiry (Ch. 3, n. 2 above), II:218; Middleton, 57. (3) Report (Ch. 3, n. 2 above), I:71. (4) Inquiry (Ch. 3, n. 2 above), II: 148-151; Middleton, 58. (5) Inquiry (Ch. 3, n. 2 above), II:150-155. (6) Ibid., 289, 155, 472. (7) Ibid., 256. (8) Report (Ch. 3, n. 2 above), I:70, 78.

10. The 1905 & 1906 Seasons

(1) Middleton, 57. (2) Inquiry (Ch. 3, n. 2 above), II: 396; Middleton, 69. (3) Inquiry II: 353, 521. (4) Report (Ch. 3, n. 2 above), I: 79.

11. Early Warning Signs

(1) Report (Ch. 3, n. 2 above), I: 77. (2) Middleton, 72. (3) Inquiry (Ch. 3, n. 2 above), II: 227. (4) Report (Ch. 3, n. 2 above), I: 79. (5) Ibid., 80. (6) *Le Soleil*, Sept. 14, 1907, p.1. (7) Inquiry (Ch. 3, n. 2 above), II:108-109; Tarkov, John. "A Disaster in the Making," American Heritage of Invention & Technology 1 (Spring 1986). (8) Report (Ch. 3, n. 2 above), I:81-2. (9) Ibid. (10) Ibid., 84. (11) Inquiry (Ch. 3, n. 2 above), II:321. (12) Ibid., 230. (13) Ibid., 177. (14) Report (Ch. 3, n. 2 above), I:85. (15) Inquiry (Ch. 3, n. 2 above), II:189. (16) Report (Ch. 3, n. 2 above), I:86.

12. Tuesday, August 27th

(1) Inquiry (Ch. 3, n. 2 above), II:188. (2) Ibid., 232. (3) Ibid. (4) Ibid., 259. (5) Ibid, 154-5; Report (Ch. 3, n. 2 above), I:98. (6) Inquiry (Ch. 3, n. 2 above), II:172. (7) Ibid., 259-262. (8) Ibid. (9) Ibid., 206-207.

13. Wednesday, August 28th

(1) Inquiry (Ch. 3, n. 2 above), II:232-233. (2) Ibid., 262. (3) Ibid., 282. (4) Ibid.; Report (Ch. 3, n. 2 above), I:88. (5) Inquiry (Ch. 3, n. 2 above), II:128-129. (6) Ibid., 127. (7) Ibid., 294. (8) Report (Ch. 3, n. 2 above), I:88. (9) Inquiry (Ch. 3, n. 2 above), II:107. (10) Ibid., 115 (11) Ibid., 125. (12) Report (Ch. 3, n. 2 above), I:90. (13) Inquiry (Ch. 3, n. 2 above), II:189. (14) Ibid., 209. (15) Ibid., 281.

NOTES

14. Thursday, August 29th

(1) Inquiry (Ch. 3, n. 2 above), II:161, 567. (2) Ibid., 357. (3) Ibid., 233-234. (4) Ibid., 74, 567. (5) Ibid., 312-313, 394. (6) Ibid., 312. (7) Ibid., 314. (8) Ibid., 263. (9) Ibid., 235, 265. (10) Ibid., 357. (11) Ibid., 261; Report (Ch. 3, n. 2 above), 91. (12) Inquiry (Ch. 3, n. 2 above), II: 261-2; Cooper's telegram is at p. 539 (13) Ibid., 103. (14) Ibid., 77. (15) Ibid., 199-200.

15. The Collapse

(1) Inquiry (Ch. 3, n. 2 above), II:74-8. (2) Ibid., 201. (3) Ibid., 185-7. (4) Ibid., 92-4. (5) Ibid., 96-8. (6) Ibid., 101. (7) Ibid., 207-8. (8) Ibid., 165-7. (9) Ibid., 208-10. (10) Ibid., 105-117. (11) Ibid., 156-7. (12) Ibid., 90-2. (13) Ibid., 82-4. (14) Ibid., 85. (15) Ibid., 297-8. (16) Ibid., 212-14. (17) Middleton, 80: Ulrich Barthe memoirs- Unpublished manuscript, papers of S.N. Parent, Bibliothèque et Archives Nationales du Québec (BANQ). (18) Report (Ch. 3, n. 2 above), I:95. (19) Middleton, 83.

16. The Devastation

(1) Inquiry (Ch. 3, n. 2 above), II:314, 540; Middleton, 84. (2) L'Hébreux, Michel. *Le Pont de Québec,* Les Éditions La Liberté, Sainte-Foy, QC, 1986, p.49. (3) Blanchard, David. *High Steel: The Kahnawake Mohawk and the High Construction Trade,* 1981; Manuscript in the library of the Kanien'kehaka Raotitiohkwa Cultural Center, Kahnawake, QC, pp. 13-14; Middleton, 84. (4) Inquiry (Ch. 3, n. 2 above), II:264, 357. (5) L'Hébreux, 53; Middleton, 87. (6) Ibid., 88. (7) *New York Times,* September 1, 1907, p.3. (8) *Montreal Daily Star,* September 1, 1907, p.12. (9) Ibid. (10) Report (Ch. 3, n. 2 above), I:89-90. (11) Inquiry (Ch. 3, n. 2 above), II:182. (12) Red File, Indian Affairs, (RG-10, Volume 2906, File 185-723-3). (13) Ibid. (14) Ibid. (15) Ibid. (16) Ibid. (17) Ibid. (18) Middleton, 88.

17. The Coroner's Inquest

(1) *Le Soleil*, September 3rd, 1907, p.1. (2) Transcript of the Inquest Into the Death of Zephirin Lafrance, September 3, 1907, BANQ, (hereinafter referred to as Inquest), 1. (3) Ibid., 2. (4) Ibid., 4. (5) Ibid., 9. (6) *Le Soleil*, September 3, 1907, p. 1, 6. (7) *Quebec Chronicle*, September 4, 1907. (8) Inquest, 12. (9) Ibid., 16. (10) Ibid., 17. (11) Ibid., 11. (12) Ibid., 12. (13) Ibid., 13. (14) Ibid., 22. (15) Ibid., 23. (16) Ibid., September 5, 1907, pp. 1-7. (17) Ibid., 1-12. (18) Ibid., 13-15. (19) Ibid., 16-19.

18. The Royal Commission of Inquiry

(1) Inquiry (Ch. 3, n. 2 above), II:2. (2) Copeland, Patrick and Rod Millard, editor: *Biographical Dictionary of Canadian Engineers – Henry Holgate*. (3) Middleton, 91. (4) Inquiry (Ch. 3, n. 2 above), II:3 (5) Ibid. (6) Ibid., 5. (7) Ibid., 10-14. (8) Ibid., 16. (9) Ibid., 62-63. (10) Ibid., 64. (11) Ibid., 78. (12) Ibid., 67. (13) Ibid., 73. (14) Ibid., 78. (15) Ibid.

19. The Inquest Ends

(1) Inquest, September 12, 1907, 1. (2) Ibid., 4. (3) Ibid., 6. (4) Ibid., Verdict (handwritten version). (5) Middleton, 91. (6) *Boston Herald*, September 13, 1907, 1. (7) *The Gazette (Montreal)*, September 12, 1907, 1. (8) *The Globe* (1844-1936); Sep 13, 1907; ProQuest Historical Newspapers: *The Globe and Mail*, p. 3.

20. The Inquiry Continues

(1) Inquiry, (Ch. 3, n. 2 above), II:125. (2) Ibid., 147-158. (3) Ibid., 168-171. (4) Ibid., 172-184. (5) Ibid., 83. (6) Ibid., 211; *Le Soleil*, Sept. 20, 1907, Vol. 11, No. 22, p.1. (7) *Engineering News*, vol. 58, September 19, 1907, 315. (8) Inquiry (Ch. 3, n. 2 above), II:211-2. (9) Ibid., 212-249. (10) Ibid., 249-275. (11) Ibid.,

275-291. (12) Ibid., 304-306. (13) Ibid., 306-315. (14) Ibid., 318-9. (15) *Le Soleil* Sept. 24, 1907, p.8.

21. The Inquiry Ends

(1) Inquiry, (Ch. 3, n. 2 above), II:322-332. (2) *New York Times,* Sept. 28, 1907, 2. (3). Inquiry, (Ch. 3, n. 2 above), II: 332-343. (4) Ibid., 343-358. (5) *Engineering News,* vol. 58, Oct. 31, 1907, p.473-477. (6) Middleton, 94, fn. 18. (7) Inquiry (Ch. 3, n. 2 above), II:361-365. (8) Ibid., 369-378. (9) Ibid., 385-398. (10) Ibid., 398-406. (11) Ibid., 403-406. (12) Ibid., 406-416. (13) Ibid., 418-421. (14) Ibid., 421-423. (15) Ibid., 404-405. (16) Middleton, 95.

22. The Findings of the Commission

(1) Letter: Henry Holgate to R.J. Butler, Feb. 25, 1908, RG 43, vol. 402, file 8419, Library and Archives Canada, Ottawa (Hereafter, LAC). (2) Report (Ch. 3, n. 2 above), I:9-10. (3) Ibid., 92. (4) Select Committee Report, (Ch. 3, fn. 3 above), 137-149. (5) Kranakis, Eda (U. of Ott.), "Fixing the Blame, Organizational Culture and the Quebec Bridge Collapse," *The International Quarterly of the Society for the History of Technology,* July 2004, Vol. 45, No. 3, p. 489. (6) Ibid., 499. (7) Report (Ch. 3, n. 2 above), I:83; Inquiry (Ch. 3, n. 2 above), II:396. (8) Report (Ch. 3, n. 2 above), I:109. (9) Inquiry (Ch. 3, n. 2 above), II:391. (10) Kranakis, p.513-514. (11) *Engineering News* brought attention to these issues and called them, "errors of administration." March 9, 1908, p. 318. (12) Report (Ch. 3, n. 2 above), I:35. (13) Kranakis, *510.* See also Waddell's paper on nickel steel: *The Globe* (1844-1936), May 22, 1908; ProQuest Historical Newspapers: *The Globe and Mail,* pg. 4. (14) Report (Ch. 3, n. 2 above), I:93. (15) Ibid., 90. (16) *Engineering News* March 19, 1908, p. 318. (17) Report (Ch. 3, n. 2 above), 93. (18) Ibid., 52. (19) Ibid., 92. (20) Ibid. (21) Ibid., 93. (22) Inquiry (Ch. 3, n. 2 above), II:262. (23) Ibid., 357. (24) Ibid., 413. (25) Ibid., 405. (26) Report (Ch. 3, n. 2 above), I:91. (27) Inquest, 12. (28) Inquiry (Ch. 3, n. 2 above), II:374-375. (29) Report (Ch. 3, n. 2 above), I:87. (30) Ibid., 49. (31) Ibid., 51. (32) *The Gazette,* Wed. Mar 11. 1908, Vol 61. (33) Select Committee Report, (Ch. 3, fn. 3 above), 193. (34) *Engineering News,*

Oct. 31, 1907, p. 473. (35) Report (Ch. 3, n. 2 above), I:49. (36) Ibid., 75. (37) Ibid., 50. (38) *The Engineering Record*, March 21, 1908, vol. 59, no. 12, p. 331. (39) Middleton, 63 (40) 12770 House of Commons, July 13, 1908.

23. The Aftermath

(1) Letter: John McMahon to Prime Minister Laurier, 10 October 1908, Wilfrid Laurier Papers, Correspondence (microfilm), reel C-867, p.145805, LAC. The newspaper article (from the *Toronto Globe*) was included with McMahon's letter. (2) Kranakis, 517. (3) *Engineering News, vol.* 60, September 17, 1908, p. 307. (4) Select Committee Report, (Ch. 3, fn. 3 above), 167, 172. (5) Special Despatch to *The Globe* (1844-1936); Jul 18, 1908; ProQuest Historical Newspapers: *The Globe and Mail*, p. A1. (6) Proceedings of the House of Commons July 17, 1908, p.13395; Ref. Senate debates re Parent - friend of Laurier p. 1726. (7) *Engineering News, vol.* 60, September 17, 1908, 307. (8) *The Gazette*, Montreal, September 12, 1916, pp. 1, 8; *The Globe*, Toronto, September 12, 1916, pp. 1, 3. (9) *The New York Times*, August 25, 1919. (10) Middleton, 99. (11) Letter: K.C. Stuart to John Deans, August 1, 1908 – report of settlements with families of deceased. (12) Report, Board of Engineers, Quebec Bridge, November 20, 1918, letter C.N. Monsarrat to Major Bell, Acting Deputy Minister, Dept. Railways and Canals.

CPSIA information can be obtained
at www.ICGtesting.com
Printed in the USA
LVHW100745260422
717138LV00006B/417